Barts' Guide To Algebra

Barts' Guide To Algebra

Will Duncombe

AuthorHouse™
1663 Liberty Drive
Bloomington, IN 47403
www.authorhouse.com
Phone: 1-800-839-8640

Published by AuthorHouse 01/29/2013

ISBN: 978-1-4772-3875-2 (sc)
ISBN: 978-1-4772-3876-9 (e)

The Barts' Guide To.. Algebra!

Will Duncombe, October 14th 2012

TABLE OF CONTENTS

INTRODUCTION

Algebra is really just a kind of language. But it is actually one that you can learn the basics of more easily than most.

Part 1 of this chapter mostly shows you the grammar (i.e. the rules) of the language of algebra. It should equip you with what you need to know in order to do later things, like solving equations. It does take a while to build up a solid foundation to algebra, and this can sometimes be quite tedious, but do just persevere, even if sometimes you wonder where it's all going.

The second part of the chapter, on applying algebra, is about doing things using algebra. These are the bits that earn you lots of marks in exams. But you won't be able to do them properly or confidently unless you've taken in Part 1 first.

My aim is that, by the end of this chapter, you'll be able to find some pleasure in algebra. That may sound hard to imagine right now but, if you're willing to persevere, you may be in for a surprise !

How To Use This Chapter

Although, as I write this, I'll be thinking about those of you who want to take first exams in maths, and perhaps go further, this chapter is for anyone who wants to learn algebra. I'll assume that you know absolutely nothing about the topic to start with, so just work through everything steadily and you should build up enough knowledge to deal with each new topic, as it arises. Just read the text, look at the examples and do the exercises (all of them).

Where there is space available, I suggest you write down the answers on the pages **in pencil**. Otherwise use loose paper. Check each individual answer, **as you go**, against the answers at the back, and don't move on until you understand each one. When you come to tests, however, complete the whole test before you look at the answers. Then revise and repeat them, if necessary, before you move on.

Questions marked with a * are a bit trickier. Don't miss these out, however. Have a go at them and give yourself a pat on the back if you get them right. Those marked ** really are a step higher. Don't worry if you get them wrong but, if you get them right, ... celebrate! If you get a lot of * and ** questions right, it means that you're probably good enough to go on to more advanced maths, if you decide you want to. Good luck!

Before You Start

In order to get on with algebra, there are some arithmetical things that you need to feel comfortable about first. You should really spend time on these, if necessary, before you start algebra proper.

Times Tables

I cannot emphasize enough that you really need to know your times tables, from ones to tens. You need to know them by rote - being able to work them out isn't good enough. Instant recall is what's needed. Yes, it's quite hard work, i.e. boring and time-consuming, but it's hardly difficult and will be time well spent. If you don't know your tables, then when doing algebra, you'll keep having to stop to get out the calculator, and this will get in the way of your thinking about the work in hand. **Do not move on before you have your tables perfected.**

Directed Numbers

The other thing that you need to be comfortable with is what are called 'directed numbers'. That's all to do with addition, subtraction, multiplication and division of numbers, which may be positive or negative.

So, before you start algebra, do the directed numbers test that follows (test 1). If you get on OK with this, you should find that algebra will fall into place. If not, you need to sort out your directed numbers first.

TEST 1 Time allowed: about 10 minutes. Score one mark per question or per part of a question, unless stated otherwise. Show all necessary working. Do not refer to the answers at the back until you have finished the test. **No calculators !**

You may wonder why some of the negative numbers below are in brackets. This is purely a convention (a standard way of doing things). Mathematicians often put negative numbers in brackets when they are multiplied or divided by other things, or when some 'operation' signs (+, - , x, ÷) occur immediately next to each other. Don't worry about the convention for now. I will guide you when I think you might need it.

Work out the values of the following:

1) $2 + 6$

2) $2 - 6$

3) $-2 + 6$

4) $-2 - 6$

5) 2×6

6) $2 \times (-6)$

7) $(-2) \times 6$

8) $(-2) \times (-6)$

9) $6 \div 2$

10) $6 \div (-2)$

11) $(-6) \div 2$

12) $(-6) \div (-2)$

13) $2 + 3 + 4 + 5$

14) $-2 - 3 - 4 - 5$

15) $-2 + 3 - 4 + 5$

16) $-6 - (-7)$

17) $(-3)^2$ — $(-3)^2$ means (-3) times (-3).

18) $20 + (-10) - 4 - (-6)$

19) $(-4)(-2) - (-3)$

(-4)(-2) means (-4) times (-2).

20) $\frac{(-4)(-2)(-3)}{(-6)}$

Multiply out the top, then divide it by the bottom.

Now look up the answers in the back. Unless you got all or nearly all right, you **must** practise more before starting algebra. If you try to proceed with algebra without being comfortable with directed numbers, believe me, you're heading for **big** trouble, and the algebra will feel much, **much** more difficult.

PART 1: ALGEBRAIC MANIPULATION

Algebra is really just like arithmetic, but using symbols as well as numbers. The symbols are usually letters of the alphabet, but you can use any character or shape that you like. In some ways you should find algebra easier than arithmetic, especially if your times tables aren't what they should be !

Section 1: THE USE OF SYMBOLS

Algebra can be used as a form of shorthand, whereby statements involving numbers can be written down in abbreviated form. The following examples show how such written statements can be expressed in the language of algebra, using symbols.

Example 1.1

The sum of two numbers means the two numbers added together. If the numbers are p and q, then their sum can be expressed as $p + q$.

Example 1.2

The difference between two numbers means the bigger one minus the smaller one. If the numbers are m and n, and if m is the bigger, the difference can be expressed as $m - n$.

Example 1.3

Four times a number. If the number is r, then this can be written as $4 \times r$. By convention in algebra, we miss out the times sign. Thus $4 \times r$ is written as 4r.

Example 1.4

A number divided by 5. If the number is p, then p divided by 5 can be written as

$p \div 5$, or 0.2p or $\frac{p}{5}$, or $\frac{1}{5}p$

Example 1.5

One number, s, divided by a second number, t, can be written as $\frac{s}{t}$.

Example 1.6

The **product** of two or more numbers means those numbers multiplied together (**not** added together). You must try to remember that. The product of three numbers, a, b, and c, can therefore be written as axbxc, that is, abc (remember, we omit the times signs). It is very important that you remember that abc is the same as acb and bac and bca and cab and cba. The order in which we multiply two or more numbers makes no difference to the outcome. The product is always the same. You can use this to simplify multiplications. For example

5x9x2 is the same as 2x5x9, i.e. 90. Notice that the second way is easier than the first, because multiplying by 10 is easy.

Example 1.7

3 times the product of p and q can be written as 3pq.

Example 1.8

3 times a number, u, minus 5 times a number, v, can be written as 3u - 5v.

Example 1.9

Two numbers, p and q, added together and the result multiplied by 3 can be written as (p + q)x3. By convention, we would normally write this as 3(p + q).

Note that, in the last example, we used a bracket to say that we **first** added p and q together and **then** multiplied the **result** by 3.

Exercise 1

Show all necessary working, in pencil, in the spaces available. Refer to the answers at the back after **each** question and correct your work as you do so.

Translate the following into algebra.

1) 6 times a number, p.

2) 4 times a number, y, then add 5.

3) Twice a number, n, minus a second number, c.

4) 5 times the product of a, b and c.

5) Half of a number, d. (Note that this can be written in a number of different ways, one with a decimal preceding d, another with a fraction preceding d, another as a single fraction with d on the top and another using a ' ÷' sign. See how many of these you can think of).

6) 3 times a number, p, minus 4 times a number, q.

7) The sum of two numbers, a and b, and the result multiplied by 4 (tip: use brackets).

8) The difference between two numbers, a and b, and the result divided by 3. Assume a is bigger than b.

9) Two numbers, c and d, added up and the result divided by g.

* 10) Two numbers, c and d, added together and the result multiplied by itself (tip: use brackets).

* 11) 2 times the sum of two numbers, a and b, minus 3 times the difference of the same two numbers, where a is bigger than b (tip: you'll need to use two sets of brackets).

For the following questions, translate into algebra. You'll need to use an 'equals' sign (=) somewhere in each of these. That is, you will be writing out 'equations'. Do not attempt to 'solve' the equations, just write them out.

12) 3 times a number, a, added to 4 times another number, b, all add up to 10.

13) The angles of a triangle are a, b and c. The sum of the angles of the triangle is 180 degrees.

* 14) Two numbers, a and b, are added together. The result is then multiplied by 4, making a total of 100 (use brackets).

* 15) The average (mean) of four numbers, a, b, c, and d is 8 (tip: to get the mean of a set of numbers, you add them up and divide the result by how many numbers there are).

Four Conventions That You Should Know About

1) You should have read previously that, whilst it's not actually wrong to write

$(a + b)4$ to represent a plus b added together and the result multiplied by 4, it's more conventional to put the number in front of the bracket, i.e. $4(a + b)$.

2) If the first of a string of numbers or letters has no negative sign in front of it, then we assume that it is positive. That may sound a bit obvious, but it does actually need to be stated. For example, the expression,

$2 + 3 + 4 - 5$

does actually mean

$+2 + 3 + 4 - 5$ (not $- 2 + 3 + 4 - 5$)

3) Whilst we write 2a, 3b, 6mn, etc, we omit the 1 when intending 1a, 1b, 1mn, etc. It isn't actually wrong to write 1r, for example, (and indeed you may often find it helpful to write in the figure 1) but, by convention, we normally write simply r, when meaning one lot of r.

4) If you have a string of numbers with 'operation' signs (+, - , x, ÷) between them, you must do the multiplications and divisions before the additions and subtractions. For example,

$$2 \times 3 + 4 = (2 \times 3) + 4 = 6 + 4 = 10$$

Similarly,

$$12 + 15 \div 5 = 12 + (15 \div 5) = 12 + 3 = 15$$

Section 2: SUBSTITUTION

The process of finding the numerical value of an algebraic expression, for given values of the symbols that appear in it, is called substitution. To do this, as the title suggests, you just re-write the expression, but substituting the numerical values for the appropriate letters. Then you find the value of the result.

As stated previously, multiplication signs in algebra are usually omitted. For example, 2 times y is written as 2y. The missing multiplication signs must be replaced when the letters are substituted with numerical values. That is, if y is 5, 2y is 2 x 5, i.e. 10 (not 25).

Examples

If $a = 3$, $y = 4$ and $z = 5$, evaluate (i.e. find the values of):

Example 2.1

$$2y + 3$$

Answer: $2y + 3 = 2 \times 4 + 3 = 8 + 3 = 11$

Note that, according to convention 4 of the previous section, we had to multiply the 2 and 4 together first, before adding the 3. If we did the addition before the multiplication, we'd get 14, rather than 11.

Example 2.2

$$3y + 5z$$

Answer: $3y + 5z = 3 \times 4 + 5 \times 5 = 12 + 25 = 37$

Example 2.3

$$8 - a$$

Answer: $8 - a = 8 - 3 = 5$

Example 2.4

$$\frac{3y + 2z}{a + z}$$

Answer: $\frac{3y + 2z}{a + z} = \frac{3 \times 4 + 2 \times 5}{3 + 5}$

$= \frac{12 + 10}{8} = \frac{22}{8} = 2\frac{3}{4}$

Exercise 2

If $a = 2$, $b = -3$ and $c = 5$, find the values of:

1) $c - 3$

2) $8 - b$

3) $6c$

4) $5a - 2$

5) $-c - 3$

6) $4a + 6b$

7) $\frac{ab}{8}$

8) abc

9) $4c + 6b$

* 10) $\frac{3a - 2b + c}{a - b + c}$

Work out the top. Work out the bottom.
Divide the top by the bottom.

Section 3: POWERS

The quantity 'b times b times b', or b x b x b, or bbb, by algebraic convention is written as b^3. b^3 is usually called 'b cubed' but may also be called 'b to the power 3'.

This convention is purely to save us time and space. For example, I think you'll agree that, if you have a number, c, multiplied by itself 10 times, it is so much easier to write it out and also to read it as c^{10} ('c to the power 10') than cccccccccc. There is nothing to be scared of in powers (not yet anyway !). Just try to remember that they are just there to help.

So, in the expression d^4 (spoken as 'd to the power 4') the number four represents the number of ds that are multiplied together. It is called the 'index' (plural 'indices').

So d^4 = dddd (all multiplied together)

and y^5 = yyyyy (all multiplied together).

i.e. $5^3 = 5 \times 5 \times 5 = 125$

and $3^4 = 3 \times 3 \times 3 \times 3 = 81$

Note that, when multiplying out 3 x 3 x 3 x 3, we don't have to say 3 times 3 is 9; 9 times 3 is 27; 27 time 3 is 81. We can multiply in any order we like. So here, a nice little short cut would be to say 3 x 3 x 3 x 3 equals 9 x 9, which is 81. Look out for short cuts like this.

When dealing with an expression like:

$8mn^2$

it is important to appreciate that **only** the symbol 'n' is 'powered' (not the m or 8). So,

$8mn^2$ means 8mnn (all multiplied together.)

Similarly in, $2a^2b^3$

only the 'a' is 'squared' and only the 'b' is 'cubed', so,

$2a^2b^3$ means 2aabbb (all multiplied together)

However, if we use brackets:

$(2pq)^2$ means 2pq multiplied by 2pq, which is

2pq times 2pq, which is 2pq2pq. This is the same as

2 times 2 times ppqq which is $4p^2q^2$

So $(2pq)^2$ expands to $4p^2q^2$.

Examples

If a = 2, b = -3 and c = 4, find the values of:

Example 3.1

$a^2 + b^2$

Answer: $a^2 + b^2 = 2^2 + (-3)^2$

Note that the brackets here are important.
The brackets give + 9 here, = not - 9.

$4 + 9 = 13$

Example 3.2

$(a + b + c)^3$

Answer: $(a + b + c)^3 = (2 + (-3) + 4)^3$

$= (2 -3 + 4)^3 = (3)^3 = 3 \times 3 \times 3 = 27$

Example 3.3

$2a^2b^3$

Answer: $2a^2b^3 = 2 \times 2^2 \times (-3)^3$

$= 2 \times 4 \times (-3) \times (-3) \times (-3)$

$= 2 \times 4 \times (-27) = -216$

Exercise 3

If a = 2, b = -3 and c = 4, find the values of:

1) a^2

2) b^4

3) $5a^3 + 6b^2$

4) $\frac{3a^4}{c^2}$

* 5) c^a

This isn't difficult. It just looks it!

* 6) a^{-b}

If you got the previous question right, you can get this one too!

Section 4: 'TRIAL & IMPROVEMENT' TO SOLVE EQUATIONS

NB You will need to use a calculator **with bracket keys** for this section. If you haven't used brackets on a calculator before, have a look at the instructions. It is quite easy, when you get used to it.

Technically, this section belongs in Part 2, rather than Part 1, of this chapter, but it's good practice for substitution and for powers. It'll also get you used to the idea of using brackets on your calculator, so it actually fits in quite nicely here.

Firstly, here's a brief introduction to equations. An equation is a statement that two expressions are equal. For example:

$x + 2 = 7$

You probably don't need me to tell you that the **solution** of this equation is that x must be 5 ($x = 5$). This is because the number 5 is the (only) value of x that will make the equation true (because $5 + 2 = 7$).

There are various ways to solve equations. In the above case, the solution was found by 'inspection' (i.e. just looking at it).

Another way to solve an equation is to guess a solution and then substitute it, to see if it 'works'. For example, supposing we want to solve the equation,

$x^2 + 2 = 6$

If we substitute the number 1 for x, we get $1^2 + 2$, which makes 3. So, $x = 1$ doesn't 'work' and so $x = 1$ is **not** a solution of the equation. It looks like we need a bigger number.

Let's try 2. If we substitute x by 2 in the left hand side of the equation, we get $2^2 + 2$, which **is** 6. So $x = 2$ 'works'. So $x = 2$ **is** a solution to the equation $x^2 + 2 = 6$ (because $2^2 + 2 = 6$).

This process is called 'trial and improvement'. It can be a very cumbersome method of solving an equation, unless you 'get lucky' early on and it has other disadvantages as well. However, it is sometimes a good way to find a solution to equations that contain x^3, x^4, or a higher power of x.

'Cubic' equations are equations in which the highest power of x is 3. For example:

$x^3 + 3x = 30$ is a cubic equation.

Let's try and find a solution for this equation by trial and improvement, giving the answer correct to one decimal place, if necessary. So, we're looking for a value of x which 'works' for the equation,

$x^3 + 3x = 30$

That is to say, we're looking for a number which, when cubed and then added on to three times itself will give us 30.

Well, it's not going to be a big number, as cubes 'grow' very quickly ($2^3 = 8$, $3^3 = 27$, $4^3 = 64$, etc.).

So let's start by seeing if $x = 2$ is close to being a solution of the equation. Substituting 2, in place of x, in:

$x^3 + 3x$ we get,

$(2.00)^3 + 3(2.00)$ meaning 2 cubed plus 3 lots of 2.

Input the numbers and brackets in **exactly** the format shown above on your calculator. That is, include two decimal places. You'll see why soon. When you press the 'equals' button on the calculator, the output will be '14' because,

$(2)^3 + 3(2) = 8 + 6$, which is 14.

So, $x = 2$ gives a value which is less than 30. So $x = 2$ is not a solution, but it looks like there is a solution which is bigger than 2. Let's try 3. Again, use brackets on the calculator. Notice also that you should be able to 'overtype' onto the previous expression on your calculator, so this can save quite a lot of time.

$x^3 + 3x = (3.00)^3 + 3(3.00) = 27 + 9 = 36$

This gives a value which is bigger than 30. So $x = 3$ is not a solution, but it looks like there is a solution which is less than 3. It would seem reasonable therefore that there must be a value between 2 and 3 (that is to say a decimal number) which is a solution.

It looks like the solution is nearer to 3 than 2, so let's start with 2.7. Notice that I'm going to input 2.70 into the calculator, not just 2.7. There's a good reason for this, as you'll see again later.

$x^3 + 3x = (2.70)^3 + 3(2.70) = 27.7$ . . . Too small; try 2.8.

$x^3 + 3x = (2.80)^3 + 3(2.80) = 30.3$ . . . Too big.

So, if 2.7 is too small and 2.8 is too big, the solution lies **between** 2.7 and 2.8. So, correct to one decimal place, the solution must be is either $x = 2.7$, or 2.8.

It looks like 2.8 is going to be the solution, correct to 1 decimal place. However, you do need to **prove** this. So, to be **absolutely** sure, we see what happens when we substitute 2.75 into the equation (the number exactly half way between 2.7 and 2.8). We get:

$x^3 + 3x = (2.75)^3 + 3(2.75) = 29.0$ . . . Too small.

So, the solution is bigger than 2.75. This means that the solution must be closer to 2.8 than 2.7 and therefore, the solution, correct to one decimal place, is 2.8. Notice that you must **not** write 2.80 now, as your answer is required correct to one decimal place and must therefore must only include one decimal place.

In fact, you can solve almost any equation using trial and improvement but, as you can see, it is a bit long-winded (and can get much more complicated). So we tend to use this method only when we don't know of a better way.

Anyway, the system here is:

Procedure For Solving Cubic Equations, By Trial & Improvement (correct to one decimal place)

1) If necessary, rearrange the equation into the form:

   x-cubes, x-squares, xs = a number.

2) Write out a table like this one. (This would be the table for the example given above.)

x	$x^3 + 3x = 30$?	Too big / small

3) Guess a solution for x (or use one suggested in the question) and enter it in the first column. Use a whole number to start with.

4) If this gives a result which is too big, try a smaller whole number. If it's too small, try a larger one.

5) Continue until you have two consecutive whole numbers, one which is too big and one too small.

6) Use the next row to write the words, "Therefore a solution lies between . . . . and . . . ."

7) Now guess a solution correct to one decimal place and repeat as above until you have two consecutive one-decimal numbers, one of which is too big and the other too small. Use the next row to write the words, "Therefore a solution lies between . . . . and . . . ."

8) Now try the number that is exactly half way between the last two one-decimal place numbers. For example if these were 4.6 and 4.7, use 4.65.

9) Conclude therefore which of the two numbers in (7) is the solution, correct to one decimal place.

Remember, **use the brackets on your calculator.** Also, use two decimal places from the start (for example, input 4.00 and 5.00, rather than 4 and 5). That way you can use the left and right arrows on the calculator and overtype the previous value, rather than having to re-input the whole expression each time.

Example 4.1

Show that a solution of the equation

$$x^3 - 5x = 60$$

lies between $x = 4$ and $x = 5$. Find that solution, correct to one decimal place.

Answer

Firstly, you have to do as instructed, which is to substitute $x = 4$, then $x = 5$ into the equation and put the results in the table. Use up the first two lines for this. Since $x = 4$ gives a result that is less than 60 and $x = 5$ gives a result which is greater than 60, you may conclude that a solution lies between 4 and 5. It is best actually to write this on the next line of the table. So,

x	$x^3 - 5x = 60$?	Too big / small
4.00	44	too small
5.00	100	too big
Therefore a solution lies between 4 and 5		
4.30	58.0.....	too small
4.40	63.1.....	too big
Therefore a solution lies between 4.3 and 4.4		
4.35	60.5.....	too big
Therefore 4.3 is closer than 4.4		
Therefore the solution is 4.3 (1d.p.)		

As the result for 4.35 is too big (regardless that it is very close) the solution must be closer to 4.3 than 4.4. The solution, correct to 1 decimal place is therefore $x = 4.3$ (not 4.30, as the question asks for 1 decimal place, not two). Notice that we did not need to give lots of decimal places when we calculated the results of our guesses, only enough to show the result was too big or too small.

Example 4.2

Use the method of trial & improvement to solve the equation,

$$x^3 + 2x^2 = 100$$

Answer

x	$x^3 + 2x^2 = 100$?	Too big / small
3.00	45	too small
4.00	96	too small
5.00	175	too big
Therefore a solution lies between 4 and 5 (It looks nearer 4?)		
4.20	109.3 ...	too big
4.10	102.5 ...	too big
Therefore a solution lies between 4.0 and 4.1		
4.05	99.2 ...	too small
Therefore 4.1 is closer than 4.0		
Therefore the solution is 4.1 (1dp)		

As 4.1 gives a result closer to 100 than 4.0 does, the solution, correct to one decimal place is,

$x = 4.1$

Notice, again, we don't put 4.10

Exercise 4

1) Show that a solution of the equation

$x^3 + 2x = 25$

lies between $x = 2$ and $x = 3$. Then use the method of trial and improvement to find that solution, correct to one decimal place.

x		

2) Use the method of trial & improvement, and the blank table which follows, to find a positive solution, correct to one decimal place, to the equation,

$x^3 - 4x^2 = 8$

Tip: Start with 5.00 and input into the calculator using brackets, i.e. $(5.00)^3 - 4(5.00)^2 = \ldots$

Section 5: SUMMATION OF ALGEBRAIC TERMS

Summation means the **totalling** of a string of two or more quantities (each of which may be positive or negative). You will know this as adding and subtracting, but actually subtracting may in fact be considered as the addition of a negative amount. More of that another time perhaps. Here are a couple more definitions.

Coefficients

The word 'coefficient' is very useful, and you should familiarise yourself with it. For now, I would like you to consider that the numerical coefficient of an algebraic expression is simply the number part of that expression, **including the sign**. For example, in the expression,

$2x^3 - 3x^2 + x$

The coefficient of x^3 is +2.
The coefficient of x^2 is - 3.
The coefficient of x is +1 (not 0, because x means $1x$).

Like Terms

'Like terms' are terms that are identical in every respect apart from the numerical coefficient. For example:

7p, 3p and -2p are all 'like terms' because each term is a certain number of ps. Only the numerical coefficients differ.

2ab, -3ab, -4.5ab and 21ab are all like terms because each term is a certain number of abs. Only the numerical coefficients of ab differ.

$3pq^2$, $-30pq^2$, $- pq^2$, $1.5pq^2$ and $\frac{1}{4}pq^2$ are all like terms, because each term is a certain number of pq^2 s. Only the numerical coefficients of pq^2 differ.

But, $3pq^2$, $3pq^3$ and $3pq^4$ are **not** like terms. Their numerical coefficients are the same and each term contains a p, but their powers of q are different.

An expression which contains like terms can be simplified (generally meaning reduced or abbreviated) for example:

7a + 3a - 5a simplifies to 5a.

This is because we have 7 lots of a, plus 3 lots of a, minus 5 lots of a, which make 5 lots of a. Similarly:

$3b^2 + 7b^2$ is the same as $10b^2$ because

3 lots of b^2 and 7 lots of b^2 make 10 lots of b^2.

Similarly, $2pq + 3pq - 10pq = -5pq$ (minus 5 lots of pq).

Only like terms can be simplified. Thus

$7a + 3b - 2c$

cannot be simplified as none of the terms are like terms, but

$7a + 3b - 2a$ simplifies to $5a + 3b$.

You can have several sets of like terms in an expression. These can then be gathered together and simplified. For example:

$8a + 3y + 2a^2 - a + 4y + 3a^2 - 5 + 10y$

$= (8a - a) + (3y + 4y + 10y) + (2a^2 + 3a^2) - 5$

$= 7a + 17y + 5a^2 - 5$

You can add and subtract like terms in any order you like. It is conventional, however, where straightforward, to write expressions with the higher powers first and to write any purely numerical terms last.

It is important that you appreciate that you can rearrange terms in an expression, as long as you move the terms **together with the signs that precede them**.

Thus, the above simplified expression, $7a + 17y + 5a^2 - 5$, though not incorrect, might more conventionally be written as:

$5a^2 + 7a + 17y - 5$

Examples

Simplify the following expressions as far as possible.

Example 5.1

$3mn + 5mn + mn - 6mn$

Answer: $3mn$ (plus 9 lots of mn minus 6 lots of mn)

Example 5.2

$pq - 4pq + 6 - 2pq - 10$

Answer: $-5pq - 4$ (plus 1 lot of pq, minus 6 lots of pq, plus 6, minus 10)

Example 5.3

$ab + 6ab^2 - 3ab^3 + 4ab^2 - 5ab^3$

Answer: $- 8ab^3 + 10ab^2 + ab$

Exercise 5

Simplify the following expressions as far as possible.

1) $a + a + a - b - b$

2) $10b - 3b$

3) $-5c - 4c$

4) $-2d + 3d$

5) $6b - 3b + 8b - 4b + b$

6) $-2c - 3c - 5c + 20c$

7) $7pq - pq - pq - 2pq$

8) $10xy + 2xy - xy$

9) $3r^2 + 5r^2 - 10r^2$

10) $3m - 4n + 5n - 3m - 2n + 3m$

11) $x^2 + 3x - 5x - 2$

12) $ab^2 - 3ab + 4ab^2 - 3 + 7ab^2 + 10$

* 13) $4a^2b + 3ab^3 + 6b^4 + a^2b - 2ab^3 - 2ab$

14) $1.5a + 3.5b + 2.5a - 2b$

* 15) $2p^{27} + 12p^{27}$

Section 6: MULTIPLYING ALGEBRAIC TERMS

Although, in algebra, we can only **summate** (add and subtract) **like** terms, we can multiply and divide anything.

As seen before, when we multiply different symbols, the result is simply written as those symbols in a row. By convention (this time a very important one) when doing this, we put the sign first, followed by any numerical component, followed by the letters in alphabetical order (where possible). For example:

b times c times a (i.e. b x c x a) is written 'abc'.

When multiplying expressions which have letters in common, we use powers, rather than long strings of letters. For example:

accaacad x (-3)

= (-3)aaaacccd = $-3a^4c^3d$

I mentioned it before, but because it's so important I'll say it again: It is very important to appreciate that

abc is the same as bca and cba, etc., just as

2 x 3 x 4 = 4 x 3 x 2 = 3 x 2 x 4 (all = 24)

Similarly, a^4c^3d, dc^3a^4, a^4dc^3, c^3a^4d, etc. are all the same because they all mean aaaacccd. Whilst none of these formats is actually incorrect, always give expressions like this with the individual letters in alphabetical order, and use powers, i.e. a^4c^3d. It makes life much easier.

Multiplying Positives And Negatives

The rules are just the same as in ordinary arithmetic. That is:

A plus times a plus makes a plus.
A minus times a minus makes a plus.
A plus times a minus makes a minus.

In other words, when **multiplying** (or dividing) two numbers:

The same signs (both plus or both minus) make a plus.
Different signs (a plus and a minus) make a minus.

For example:

(+) p times (+) q = (+) pq and (-p) times (-q) = (+) pq **BUT**:

(-p) times (+) q = - pq and (+) p times (-q) = - pq

It seems obvious that any number of pluses multiplied together make a plus. However, when multiplying minuses together, an even number of minuses make a final plus, whilst an odd number of minuses result in final a minus. For example:

$(-p) \times (-p) \times (-p) \times q \times q = -p^3q^2$ (3 minuses make a minus)

but:

$(-p) \times (-p) \times (-p) \times (-q) \times q = (+)\ p^3q^2 = p^3q^2$ (4 minuses make a plus)

In summary, all multiplication in algebra should be carried out as follows.

Procedure For Multiplying Algebraic Terms

1) Multiply out the signs.

2) Multiply out the numbers.

3) Multiply out the letters, in alphabetical order, using powers if appropriate.

Example 6.1

$4m \times 5m = 4 \times m \times 5 \times m$

$= 4 \times 5 \times m \times m = 20m^2$

Example 6.2

$2pq \times (-3q^2) = 2 \times (-3) \times p \times q \times q \times q$

$= -6pq^3$

Example 6.3

$(-2bc^2) \times (-5b^3c)$

= (-2) times (-5) times bbbbccc

= (+) 2 times 5 times bbbbccc

$= 10b^4c^3$

Exercise 6

Simplify (i.e. express the following in its simplest form)

1) $3z \times 4y$

2) $2a \times 3b$

3) $4 \times 5m$

* 4) $\frac{1}{2} q \times 16s$ — Not difficult really!

5) $z \times (-y)$

6) $(-4n) \times (-3p)$

7) $6d \times (-3e)$

8) $(-3a)^2$ — Means (-3a) times (-3a)

9) $(-4e) \times 5t$

10) $2p \times 3r \times (-4q)$

11) $3w \times (-4v) \times (-u) \times 5u$

12) $(-b) \times b$

13) $(-b) \times (-b)$

14) $(-2ab) \times (-3bc)$

15) $8mn \times (-2\, m^3n^4)$

16) $4ab \times (-3b^2)$

17) $2q^3r^4 \times 5q^2r^2$

18) $2a^2 \times 3q^2$

19) $(-q^3) \times (-r^2) \times (-pqr)$

* 20) $(-3m^3)^3$ — meaning $(-3m^3) \times (-3m^3) \times (-3m^3)$

Section 7: DIVIDING ALGEBRAIC TERMS

When dividing algebraic expressions expressed in the form of a fraction, cancellation between top and bottom of the fraction may be possible. Cancelling is generally only possible, however, when all the individual terms on the top of a fraction are separated by multiply signs and all of the terms on the bottom are also. Remember that cancelling is the equivalent of dividing the top and bottom of the fraction by the same thing - and therefore does not change the value of the fraction. It is only (generally) possible to cancel therefore when all the operation signs are multiplications.

For example:

$$\frac{\cancel{p}q}{\cancel{p}} = q$$

(because p times q, divided by p, just leaves q. Try it with real numbers)

but you **cannot** cancel

$$\frac{p+q}{p}$$

(because of the + sign. Again, try it with real numbers).

The same rules for signs apply for division as do for multiplication (actually, because division is a type of multiplication). In other words, when **multiplying <u>or</u> dividing** two numbers:

Same signs (both plus or both minus) make a plus.

Different signs (a plus and a minus) makes a minus.

Example 7.1

$$\frac{-3a}{2b} = -\frac{3a}{2b}$$

(no cancelling was possible here, but notice that the sign's position changes to show the overall result is negative)

Example 7.2

$$\frac{-3a}{-2b} = \frac{3a}{2b}$$

(overall result becomes positive)

Example 7.3

$$\frac{3p^2q}{6pq^2} = \frac{{}^1\cancel{3}\ \cancel{p}p\ \cancel{q}}{{}_2\cancel{6}\ \cancel{p}\ \cancel{q}q} = \frac{p}{2q}$$

Example 7.4

$$\frac{-18ru^2v^2}{6r^2uv^5} = -\frac{{}^3\cancel{18}\ \cancel{r}\ \cancel{u}u\ \cancel{vv}}{{}_1\cancel{6}\ \cancel{r}r\ \cancel{u}\ \cancel{vv}vvv} = -\frac{3u}{rv^3}$$

Procedure To Divide One Algebraic Expression By Another

To divide a by b (i.e. a ÷ b):

1) Write down the resulting sign.

2) Express the division as a fraction, $\frac{a}{b}$.

3) Simplify where possible by cancelling.

Laws Of Indices

From the above and the previous section, you may have noticed two of the laws of powers (or indices). These are that:

1) $5^6 \times 5^3 = 5^9$ So,

$n^6 \times n^3 = n^9$ or, more generally,

$$\underline{\mathbf{n^a \times n^b = n^{a+b}}}$$

2) $5^6 \div 5^2 = 5^4$ So,

$n^6 \div n^2 = n^4$ or, more generally,

$$\underline{\mathbf{n^a \div n^b = n^{a-b}}}$$

You should also be aware that:

3) Any number to the power 1 is itself, for example,

$2^1 = 2$ and $10^1 = 10$ and $1.456^1 = 1.456$

4) Any number to the power zero is 1, for example,

$2^0 = 1$ and $10^0 = 1$ and $1.456^0 = 1$

There are other laws of indices, which you may come across later.

Exercise 7

Simplify (express the following in its simplest form)

1) $24p \div 12$

2) $5a \div (-6b)$

3) $(-4a) \div 7b$

4) $(-4s) \div (-4t)$

5) $6c \div 3c$

6) $6by \div 3y$

7) $8b^2y \div 2b$

8) $-12c^2d \div 3cd$

9) $8c^2de \div 2ce^2$

10) $8c^2de \div -2ce^2$

11) $-8c^2de \div -2ce^2$

12) $5a^6b^5 \div 6a^3b^7$

13) $\dfrac{2a^3 \text{ x } 3p^2}{12ab}$

* 14) $(c + 2)^2 \div (c + 2)$

You can cancel something with these two, despite the + signs, because (c + 2) represents a single number.

** 15) $(c + 2) \div (c + 2)^2$

Start by writing these down as a fraction.

TEST 2 Time allowed: about 40 minutes.

Show all necessary working in the spaces provided. Do not refer to the answers at the back until you have finished the test.

For this the first part of test, the answers are either "true" or "false". Underline which you think it is. **In the case of "false" give the correct answer.** There is one mark for each question or part question, except where stated, so make sure you get the easy ones right!

1) True / False $2a + 3a = 5a$

2) True / False $2a - 3a = -1$

<u>NOW STOP</u> ! Remember, as asked, for any question you answer as 'false' you must also work out the right answer. I say that because, for some reason, most people don't notice that bit of the instructions. Go back if you haven't done this.

3) True / False $-2a + 3a = a$

4) True / False $-2a - 3a = -5a$

5) True / False $-2a + (-3a) = a$

6) True / False $2a - (-3a) = 5a$

7) True / False $2a \times 3a = 5a$

8) True / False $2a \times (-3a) = -6a^2$

9) True / False $(-2a) \times (-3a) = -6a^2$

10) True / False $(-6a) \div (-3a) = -2$

11) True / False The sum of two numbers can be represented by the expression $a + b$.

12) True / False The expression $p - q$ can represent the difference between two numbers.

13) Truc / Falsc The **product** of 6 and p is 6 + p.

14) True / False Double a number, then add 3 can be expressed as 2n + 3.

15) True / False Three numbers added together and then the result being divided by a fourth number can be written as (a + b - c) ÷ d.

16) True / False The value of 3a + 7, when a is 3, is 16.

17) True / False The value of 4a - 3b, when a is 3 and b is 1 is, is 9.

18) True / False The value of 4a - 3b, when a is 3, and b is - 1 is 9.

19) True / False The value of 6ab ÷ 3c, when a is 2, b is 3 and c is - 3 is 4.

20) True / False aaaaa is the same as a^4.

21) True / False yyy is the same as y^3.

22) True / False a + a + a + a + a is the same as a^5.

23) True / False a^3b^2 is the same as ababa.

24) True / False The value of a^4, when a = 2 is 8.

25) True / False $5x + 8x$ is $13x^2$.

26) True / False $8x - 5x$ is 3.

27) True / False 7s - 2s is 5s.

28) True / False 15ab + 7ab - 3ab - 5ab is 14ab.

29) True / False 8a x 5a is 40a.

30) True / False $9z \times 5z$ is the same as $45z^2$.

31) True / False a^2b is the same as ba^2.

32) True / False $6a^2b^3$ is the same as $6a^3b^2$.

33) True / False $1 - x + x^2$ is the same as $x^2 - x + 1$.

34) True / False $2x - 3 - x^2$ is the same as $x^2 + 2x - 3$.

35) True / False $6n^2 \div (-3n) = 2n$

36) True / False $(-5pq^2) \times (-8p^2q) = 40p^3q^3$

37) Show that a solution of the equation $2x^3 - 3x = 25$ lies between $x = 2$ and $x = 3$. (N.B. You are **not** required to **evaluate** the solution). **NB 3 marks**

38) Use trial and improvement to find a solution, between 6 and 7, for the equation $x^3 - 3x^2 = 155$. Give your answer correct to 1 decimal place.

NB 5 marks

Section 8: EXPANDING BRACKETS

Brackets are used for our convenience, when grouping terms together.

$3(a + b)$ means

3 times $(a + b)$ which is the same as

3 lots of $(a + b)$ which means

$(a + b) + (a + b) + (a + b)$ which is

$a + b + a + b + a + b$ which is

$3a + 3b$.

So $3(a + b) = 3a + 3b$

We could repeat this with lots more examples and this would show us that, when multiplying out a single bracket, which we call 'expanding' the bracket, we multiply the term outside the bracket by each term inside the bracket, in turn. For example:

$5(2a + 3b)$

$= 5 \times 2a + 5 \times 3b$

$= 10a + 15b$

Working from first principles, it would seem logical that:

$3(a - b) = 3a - 3b$ and that

$3(a + b - c) = 3a + 3b - 3c$.

Be **very** careful when you have a minus in front of a bracket. When you do, the signs of each term inside the bracket change when you expand the bracket. This is because each term inside the bracket is being multiplied by a negative. And to multiply by a negative changes a plus into a minus and changes a minus into a plus. So:

$-3(a + b) = - 3a - 3b$ whilst

$-3(a - b) = - 3a + 3b$ and

$-3(a + b - c) = - 3a - 3b + 3c$.

In particular, be **especially** careful when you have a minus, but with no number, in front of a bracket. For example:

$-(2 - a)$ really means $-1(2 - a)$ which gives, $-2 + a$

When expanding (and simplifying) expressions with strings of single brackets, first expand the individual brackets in turn. Then summate any 'like terms'. For example:

$2(2a - 3b) + 3(a + 3b)$

$= 4a - 6b + 3a + 9b$

$= 7a + 3b$

So, $2(2a - 3b) + 3(a + 3b)$ is the same as,

$7a + 3b$

Examples

Expand, and simplify where possible:

Example 8.1

$2a(3a - 2b)$

Answer: $2a(3a - 2b)$

$= 2a \times 3a - 2a \times 2b$

$= 6a^2 - 4ab$

Example 8.2

$2c(1 + b) + 3c(1 + 2b)$

Answer: $2c(1 + b) + 3c(1 + 2b)$

$= 2c + 2bc + 3c + 6bc$

$= 5c + 8bc$

Example 8.3

$2r(1 - r) - 3r(r - 1)$

Answer: $2r(1 - r) - 3r(r - 1)$

$= 2r - 2r^2 - 3r^2 + 3r$ NB: $+3r$, **not** $- 3r$

$= 5r - 5r^2$ (which is the same as $- 5r^2 + 5r$)

Example 8.4

$2n(3 - 4n) - n^2(10 - n)$

Answer: $2n(3 - 4n) - n^2(10 - n)$

$= 6n - 8n^2 - 10n^2 + n^3$ NB: $+ n^3$, **not** $- n^3$

$= 6n - 18n^2 + n^3$

It is acceptable to leave the result in this format, but often it might be 'tidied up' by re-writing in descending order of powers of n, i.e.:

$= n^3 - 18n^2 + 6n$

Again, note that we can shift terms around in any 'summation' string (i.e. any string being added and subtracted) but terms must be shifted **together with the signs that precede them**.

Exercise 8a

Expand the brackets in the following:

1) $3(a + 1)$

2) $2(3e - 4f)$

3) $p(3q + 4)$

4) $-(a + b)$

5) $-(3a - 2c)$

6) $-2p(p - 1)$

7) $a(b + c - d)$

* 8) $2a^2(a^2 + 3a - 1)$

For the following questions, expand the brackets **and** simplify:

9) $3(n + 1) + 2(n + 5)$

10) $2(a - 1) + 3(1 + a)$

11) $5(a + 1) + 4(a - 1) + 3(1 - a)$

12) $3(n - 1) - 2(n + 5)$ — be careful with the minus in front of the bracket.

13) $3(n - 1) - 2(n - 5)$ — Again, take care with the minus.

14) $3(2a - 3b) - (4a - 5b)$ — And again!

15) $2r(3r - 1) + 3(2 - r^2) + 1$

16) $4p(p - 1) - p(p - 2)$

* 17) $3(a - b) - 2(2a - 3b) + 3(3a - b)$ — The * is for accuracy, not difficulty.

Expanding Double Brackets

The expression $(a + b)(c + d)$ means:

$(a + b)$ times $(c + d)$, which is the same as:

$(a + b)$ lots of $(c + d)$.

If you think about it,

(a and b) lots of $(c + d)$ should be the same as:

a lots of $(c + d)$ and b lots of $(c + d)$, so

$(a + b)(c + d)$ is equal to $a(c + d) + b(c + d)$, that is:

$(a + b)(c + d) = ac + ad + bc + bd$.

In practice, what we do in multiplying out (i.e. expanding) a pair of double brackets is to multiply each of the terms in the first bracket by each of the terms in the second bracket. We can actually do this in any order we like, but it avoids confusion if you stick to the same order.

It can help by covering up the second term in the first bracket and carrying out the remaining expansion, then covering up the first term in the first bracket and multiplying out the other expansion.

Alternatively, you may find it helpful at first to physically connect each term in the first bracket to each in the second, using curved arcs. In the example below, draw an arc from the 'a' in the first bracket to the 'c' in the second and then another from the 'a' again to the 'd' in the second bracket. Draw these first two arcs above the brackets. Next, draw an arc from the 'b' in the first bracket to the 'c' in the second and then another from the 'b' again to the 'd' in the second bracket. Draw these two arcs below the brackets. Then multiply out the sets of terms in the same order and write them down. For example,

$$(a + b)(c + d) = ac + ad + bc + bd.$$

In this expansion, there are no 'like terms' to gather together (i.e. to simplify) but you should always check to see if you can. More often than not, you will be able to. For example:

$$(a + 7)(a + 2) = a^2 + 2a + 7a + 14$$

$$= a^2 + 9a + 14$$

I won't put in any connecting arcs, but you should feel free to do so until you feel comfortable to work without them.

Examples

Expand the following and simplify where possible.

Example 8.5

$$(n - 3)(n + 3)$$

Answer: $(n - 3)(n + 3)$ expands to

$n^2 + 3n - 3n - 9$ which simplifies to

$$n^2 - 9$$

Example 8.6

$$(2c - 3d)(3c - d)$$

Answer:

$$(2c - 3d)(3c - d) = 6c^2 - 2cd - 9cd + 3d^2$$

$$= 6c^2 - 11cd + 3d^2$$

Example 8.7

$(2p - 3q)^2$

Answer:

$(2p - 3q)^2 = (2p - 3q)(2p - 3q)$

$= 4p^2 - 6pq - 6pq + 9q^2$

$= 4p^2 - 12pq + 9q^2$

Exercise 8b

Expand and simplify, where possible:

1) $(n + 1)(n + 3)$

2) $(a + 3)(a - 2)$

3) $(2c + 3)(c + 5)$

4) $(5c + 2)(c - 4)$

5) $(3a + 1)(2a - 5)$

6) $(3n + 4)^2$

$(3n + 4)^2 = (3n + 4)(3n + 4)$

7) $(a + b)(a - b)$

8) $(2x + 3y)(3x - 4y)$

9) $(2 - 3x)(1 - 4x)$

* 10) $(a + b)(a + b + c)$ Don't give up without a fight. The theory's the same!

Section 9: FACTORISING

Common Factors

We know that:

$3(a + b)$ multiplies out to:

$3a + 3b$. This process is called expansion.

We need also to be able to do the reverse process, that is, to recognise that:

$3a + 3b$ can be expressed as:

$3(a + b)$. The process this way is called **factorisation**.

In other words, factorising is the reverse of expanding.

In arithmetic, 2 and 5 are said to be 'factors' of 10, because 2 times 5 is 10. Similarly, in algebra, 3 and $(a + b)$ are said to be factors of $3a + 3b$ because 3 times $(a + b)$ is equal to $3a + 3b$.

When trying to factorise an expression, **always** look **first** for 'common factors' (rather than other things, which we'll come across a bit later).

For example, the expression:

$2a + 6$

has the factor 2 common to each term in it. That is to say that 2 will divide exactly into both 2a and 6.

To factorise $2a + 6$, we first 'take out' the common factor, 2, that is we place the number 2 in front of a bracket. To find out what goes inside the bracket, ask yourself, "What do I need to multiply this by to get the terms of the original expression?" Thus:

$2a + 6 = 2(\ ?\ .\ .\ .\ .\ ?\ )$

To get 2a, we need to multiply 2 by a. To get 6, we need to multiply 2 by 3. So:

$2a + 6 = 2(a + 3)$

Providing you 'take out' **all** the common factors, you should always get factorising right, because you can check it by expanding. In the above, for example:

$2(a + 3)$ multiplies out to:

$2a + 6$

so the expression was correctly factorised.

Examples

Factorise completely the following:

Example 9.1

$$2c^2 - 3c$$

Answer

Firstly, c will go into both $2c^2$ and into 3c, so:

$$2c^2 - 3c = c(? . . . . ?).$$

Now, to get $2c^2$, we need to multiply c by 2c. To get 3c, we need to multiply c by 3. So:

$$2c^2 - 3c = c(2c - 3).$$

We can check this by expanding c(2c - 3), which indeed gives us $2c^2 - 3c$. So, given that c was the only common factor, the factorisation must be correct. If it were not, we'd just try again, until we do get it right.

Example 9.2

$$3m - m^2$$

Answer

$$3m - m^2$$

$$= m(3 - m)$$

Example 9.3

$$9mn^2 - 3n$$

Answer

In this case, there is more than one common factor. Both 3 and n will 'go' (i.e. divide exactly) into $9mn^2 - 3n$. So, in this case 3n is the 'highest common factor'. We thus take out 3n as the common factors and then proceed as usual. Thus,

$$9mn^2 - 3n = 3n(3mn - 1)$$

Note that the second term inside the bracket is 1, not 0 (zero). Always watch out for this. 3m times what, gives you 3m? The answer is 3m x 1 = 3m.

Now, if we multiply out 3n(3mn - 1) we do indeed get $9mn^2 - 3n$, so the factorisation is be correct.

*** Example 9.4**

$$9x^3y^2 - 6x^4y^3$$

This is a bit tougher, but the principles involved are just the same. In cases like this, look first for the **highest** common factors and take them out together. So:

3 will divide exactly into $9x^3y^2$ and $6x^4y^3$.

Not only x, but x^3, will divide exactly into $9x^3y^2$ and $6x^4y^3$.

Not only y, but y^2, will divide exactly into $9x^3y^2$ and $6x^4y^3$.

So the factor to take out is $3x^3y^2$. And thus,

$$9x^3y^2 - 6x^4y^3 = 3x^3y^2(3 - 2xy)$$

If you expand this out, you will see that it is correct.

Exercise 9a

Factorise fully:

1) 5a + 5b

2) 3a - 15

3) 4a - 2ab

4) pd - pe

5) 3a + 6b - 9c

6) $7d - 3d^2$

7) $9d - 3d^2$

8) $x^3 + x^2 + x$

9) $2x^2 + 4x$

10) $ab^2 - ab$

11) $ab^3 - ab^2$

Be careful. The common factor is not just ab. What is it?

12) $m^2n + mn^2$

13) $3m^2n + 6mn^2$

* 14) $3m^5n^3 + 6m^4n^2$

* 15) $3m^2n + 6mn^2 - 9m^2n^2$

Factorising x^2 Expressions Into Double Brackets

As factorising is the opposite of expanding and as:

$(y + 4)(y + 3)$ expands into $y^2 + 7y + 12$, then:

$y^2 + 7y + 12$ factorises into $(y + 4)(y + 3)$.

But what's the system, if there is one? Well, if we were asked to factorise anything, we should **always** firstly look for a common factor. There is no common factor in $y^2 + 7y + 12$ as there is nothing that will divide into **all three** of the terms.

In fact, it is safe to conclude then that, if $y^2 + 7y + 12$ factorises at all (by no means everything does) then it will factorise into double brackets. So:

$y^2 + 7y + 12$ factorises into (? . . . . ?)(? . . . . ?)

We just have to find out what goes into the brackets.

Well, the first thing in the first bracket and the first thing in the second bracket will multiply out to give the y^2. So each bracket probably starts off with y. That is,

$y^2 + 7y + 12$ factorises into (y . . . ?)(y . . . ?)

By the same reasoning, the last thing in the first bracket and the last thing in the second bracket will multiply out to give the +12. In which case they may both be whole numbers. So now,

$y^2 + 7y + 12$ factorises into $(y \ldots a)(y \ldots b)$

Where a and b are numbers which **multiply** out to +12.

In fact, it can be shown that, as long the coefficient of y^2 is 1 (i.e. as long as we have $(1)y^2$, rather than $2y^2$, $3y^2$, etc.) then:

$y^2 + 7y + 12$ factorises into $(y \ldots a)(y \ldots b)$

Where a and b **summate** to the y coefficient (in this case, +7).

So, the system is that, if

$y^2 + 7y + 12$ factorises, then it factorises to:

$(y \ldots a)(y \ldots b)$ where:

$a \times b$ = the final number (in this case, +12) and

$a + b$ = the coefficient of y (in this case, +7).

A little thought reveals that, in this case, a and b must therefore be +4 and +3 and so:

$y^2 + 7y + 12$ factorises into $(y + 4)(y + 3)$ That is,

$y^2 + 7y + 12 = (y + 4)(y + 3)$

It is important that you appreciate that,

$(y + 4)(y + 3)$ is the same as $(y + 3)(y + 4)$

as each bracket represents a number and those numbers multiplied together come to the same product, whichever way they are written.

In the above example, all of the terms in the expression to be factorised were positive, so the signs in the brackets both had to be positive. In many cases, however, a and/or b may be negative. To determine the signs of a and b, it is vital to include the sign of the y coefficient and the sign of the number in the expression to be factorised, when considering what a and b must multiply and summate to. First of all, here's a bit of practice.

Example 9.5

Find the values of a and b, such that:

$a \times b = -12$ and $a + b = -1$

Answer

To multiply to a negative number, one must be positive and the other must be negative. To multiply to - 12, we need either:

12 and 1, or 6 and 2, or 3 and 4 with one positive and one negative.

How can we summate - 1 from these (with a plus and a minus)? We can't get - 1 from 12 and 1, nor from 6 and 2, but we can with 4 and 3, if the 4 is negative and the 3 is positive.

So a and b are - 4 and +3.

Example 9.6

Find the values of a and b, such that:

$a \times b = +16$ and $a + b = -10$

Answer

To multiply to a positive number, they must either both be positive or both be negative. To multiply to +16, we need either:

16 and 1, or 8 and 2, or 4 and 4 with either both positive or both negative.

How can we summate - 10 from these (with both + or both -)? We can't get - 10 from 16 and 1, nor from 4 and 4, but we can with 8 and 2, if they are both negative.

So a and b are - 8 and - 2.

Exercise 9b

Find the values of a and b, such that:

1) $a \times b = +20$
$a + b = +9$

Tip: To multiply out to +, both must be + or both must be - . Then, to summate to +, both will have to be +.

2) $a \times b = -2$
$a + b = +1$

Tip: To multiply out to - , one must be + and the other - .

3) $a \times b = +6$
$a + b = -5$

Tip: To multiply out to +, both must be + or both must be - . Then, to summate to - , both will have to be - .

We can now use this method to factorise x^2 expressions.

Example 9.7

Factorise $q^2 + 7q + 10$

Answer:

$q^2 + 7q + 10 = (q \ldots a)(q \ldots b)$ where,

$a \times b = +10$: so we need 5 & 2 or 10 & 1 (both must be plus or both minus)

$a + b = +7$: so it must be $+5 + 2$.

So $q^2 + 7q + 10 = (q + 5)(q + 2)$ Note that this is the same as $(q + 2)(q + 5)$.

Example 9.8

Factorise $y^2 - 4y - 21$

Answer:

$y^2 - 4y - 21 = (y \ldots a)(y \ldots b)$ where,

$a \times b = -21$: so we need 21 & 1 or 7 & 3 (one must be plus, and one minus)

$a + b = -4$: so it must be $-7 + 3$

So, $y^2 - 4y - 21 = (y - 7)(y + 3)$ Note that this is the same as $(y + 3)(y - 7)$

Example 9.9

Factorise $y^2 - 3y + 2$

Answer:

$y^2 - 3y + 2 = (y \ldots a)(y \ldots b)$ where,

$a \times b = +2$: so we need 2 & 1 (both + or both -)

$a + b = -3$: so it must be -2 and -1

So, $y^2 - 3y + 2 = (y - 2)(y - 1)$ Note that this is the same as $(y - 1)(y - 2)$

Special Cases

The following types of expressions may be factorised in exactly the same way, but are worth a separate mention.

1: Perfect Squares

Example 9.10

Factorise $n^2 - 6n + 9$

Answer:

By the same method, this will factorise into $(n - 3)(n - 3)$. It is important to appreciate that:

$(n - 3)(n - 3) = (n + 3)^2$ so:

$n^2 - 6n + 9 = (n - 3)^2$

2: The Difference Between Two Squares

Example 9.11

Factorise $n^2 - 9$

Answer

There is no common factor so, if this factorises, it does so into double brackets. In fact, although this type of expression looks different from the ones we've just done, it isn't really. It's just that the coefficient of n is zero. That is to say:

$n^2 - 9$ is the same as $n^2 + 0n - 9$.

So we are looking for numbers, a and b, such that a and b multiply out to - 9 and summate to zero. To summate to zero, the numbers must be the same, but one must be positive and the other negative (+1 - 1, or +2 - 2, or +3 - 3, etc). Two numbers which are the same, but for the sign, and which multiply to 9 are +3 and - 3. Therefore:

$n^2 - 9$ factorises into $(n + 3)(n - 3)$.

In fact whenever you have any expression like:

$a^2 - b^2$

which is known as 'the difference of two squares', it factorises into the square root of the first plus the square root of the second, multiplied by the square root of the first minus the square root of the second. For example:

$a^2 - b^2 = (a + b)(a - b)$ and

$4a^2 - 9b^2 = (2a + 3b)(2a - 3b)$ and

$25y^2 - 1 = (5y + 1)(5y - 1)$

As stated earlier, you don't really need to worry about either of these two special cases, as you should be able to factorise them in the normal way. However, they are useful in a number of ways in higher level maths so, as I said earlier, they are worth a mention.

Exercise 9c

Factorise the following:

1) $x^2 + 8x + 7$

2) $x^2 + 6x + 8$

3) $x^2 - 7x + 10$

4) $x^2 - 11x + 30$

5) $x^2 - x - 2$

6) $x^2 - 2x - 15$

7) $x^2 + 2x - 8$

8) $x^2 + x - 12$

9) $x^2 + 8x + 16$ Final answer is $(\text{something})^2$

10) $x^2 - 6x + 9$

11) $x^2 - 16$

* 12) $1 - x^2$

* 13) $4x^2 - 1$

* 14) $9x^2 - b^2$

** 15) $x^2 + 2xy + y^2$ Don't give up just because it's a ** question. Try something and expand it to check. Try again, if necessary.

Now try the following mixed bunch. Always first look for common factors. Then consider double brackets if necessary. Watch out for perfect squares and the difference of two squares.

Exercise 9d

1) $y^2 - 4y$

2) $y^2 - 4$

3) $y^2 - 4y + 4$

4) $y^2 - 9y + 8$

5) $2y^2 - 4y$

6) $2y^2 - 4y^3$

7) $y^2 - z^2$

8) $y^3 + y^2 + y$

* 9) $25y^2 - 1$

** 10) $y^4 + 6y^2 + 9$

TEST 3 Time allowed: about 25 minutes.

1) Expand $4a(3a - a^2)$

2) Expand and simplify $2(a + b) - 3(b - a)$

3) Expand and simplify $(2a + 1)(3a + 2)$

4) Expand and simplify $(a + 3)(a - 3)$

5) Expand and simplify $(1 - n)(2 - 3n)$

6) Expand and simplify $(2a + 3b)^2$

7) Expand & simplify $(a + 2)(a + 3) + (a + 2)(a - 3)$

8) Factorise completely $6a^2b^3 - 14a^3b^5$

9) Factorise $a^2 - 5a + 4$

10) Factorise $a^2 - 25a$

11) Factorise $a^2 - 25$

12) Factorise $a^2 - a - 90$

13) Fill in the bracket: $a^2 - 10a + 25 = (\ldots .)^2$

PART 2: THE APPLICATION OF ALGEBRA

Well, part 1 was the nitty gritty of algebra. You probably found some of it quite hard, not to mention tedious. But you're still here, so well done! Part 2 now is about using what you've learned to do things with algebra.

Section 10: LINEAR EQUATIONS

As we discussed in section 4, an equation is a statement that two expressions are equal. For example:

$$a + 2 = 7$$

This is known as a **linear equation in one unknown**. It's called a linear equation because, on a graph, it would be represented by a straight line. Linear equations don't contain any powers, like a^2, a^3, a^4, etc. The equation is described as being 'in one unknown' because there is only one unknown quantity in it, 'a'.

A **solution** of an equation is a value of an unknown quantity in the equation ('a' in this case) which makes the equation true. In the case of **linear** equations, there is only **one** value of the unknown which makes the equation true (in other types of equation, there may be more than one solution, or there may actually be **no** solutions). In the case of this linear equation:

$$a + 2 = 7$$

there is therefore only one solution. That is to say, there is only one value of 'a' which makes the equation true. That value is, of course, 5. So:

$a = 5$ is the solution of the equation $a + 2 = 7$

because it is the only value of 'a' which makes the equation true, because $5 + 2 = 7$.

I can hear you say, "Well I could have told you that, and quite a meal you made of it too!" However, whilst it seems obvious in this case, it isn't always, so the above 'ramble' is worth pondering.

Unfortunately, not all equations are as straightforward as that. For example, the chances are that you wouldn't be able to tell me quite so quickly that the solution of the equation:

$$2(3a - 4) = a$$

is that 'a' is 1.6. That is, the (only) value of 'a' which makes the equation true is 1.6.

In fact, we can reduce any linear equation, step by step, until we end up with the solution, by bearing in mind the following:

1) An equation is a statement of equality between two expressions (it states nothing more than the fact that the two sides of the equation are equal in value).

2) We can add or subtract the same thing to each side of the equation and the two sides will still remain equal.

3) We can multiply or divide each side by the same amount and the two sides will still remain equal.

In fact there are other things that you can do to both sides (for example, you can square or square root both sides) but these are all you need to know about for now.

Examples

Solve the following equations.

Example 10.1

Look again at, and solve, the equation:

$a + 2 = 7$

Answer

We need to get rid of the + 2 on the left. If we do the **reverse** operation and **subtract** 2 from either side, we get:

$a + 2 - 2 = 7 - 2$	Simplifying this we get:
$a = 7 - 2$	and,
$a = 5$	The solution is $a = 5$.

Example 10.2

Solve the equation $b - 3 = 4$

Answer

We need to get rid of the - 3 on the left. If we do the **reverse** operation and **add** 3 to both sides we get:

$b - 3 + 3 = 4 + 3$	Simplifying this we get
$b = 4 + 3$	and,
$b = 7$	The solution is $b = 7$.

Example 10.3

Solve the equation $4c = 12$ (meaning 4 times c makes 12)

Answer

We need to get rid of the ‘times 4’ on the left. If we do the **reverse** operation and **divide** both sides by 4, we get:

$\frac{4c}{4} = \frac{12}{4}$ Simplifying this we get

$c = \frac{12}{4}$ (because $\frac{4c}{4}$ leaves just c)

So, $c = 3$ The solution is $c = 3$

Example 10.4

Solve the equation $\frac{d}{5} = 6$ (meaning d divided by 5 makes 6)

Answer

We need to get rid of the ÷ 5 on the left. If we do the **reverse** of this and **multiply** both sides by 5, we get:

$\frac{d}{5} \times 5 - 6 \times 5$ Simplifying this we get

$d = 6 \times 5$ (because $\frac{d}{5} \times 5$ leaves just d)

so, $d = 30$ The solution is $d = 30$

Notice that, in each example, we effectively shifted things from one side of the equation to the other, by **reversing** their operations. This fact enables us to take short cuts, which save time and energy.

Example 10.5

If $n + 2 = 3$

then $n = 3 - 2$

and $n = 1$

All we did was to shift the 2 from the left to the right. But in doing so, had to reverse its operation from plus to minus. Think of it as, you can shift something from one side of an equation to the other, that is move it across the equals sign, as long as you change plus to minus, minus to plus, times to divide or divide to times.

Example 10.6

If $n - 3 = 4$

then $n = 4 + 3$ $n = 7$

Example 10.7

If $5n = 20$

then $n = \frac{20}{5}$ $n = 4$

Example 10.8

If $\frac{n}{2} = 6$

then $n = 6 \times 2$ $n = 12$

Again, in each example, we shifted things from one side of the equation to the other, by reversing the operations that they were originally doing.

Procedure For Solving Linear Equations

Always (or to start with at least) use this precise order of events when solving linear equations. You need to make the process become second nature to you.

1) Remove denominators (by multiplying through by each of them in turn, or by 'cross multiplying').

2) Remove brackets (i.e. expand or multiply them out).

3) Collect terms. This means, if an expression contains the unknown (let's call it n) shift it to the left hand side of the equals sign. If an expression doesn't contain n, shift it to the right hand side.

4) Simplify.

5) Change the signs on both sides of the equation, if necessary. It is necessary if the n on the left has a negative coefficient.

6) Divide though by the coefficient of n.

It's important for you to become comfortable with 'collecting terms', so let's skip denominators and brackets for the moment and start with some equations where the first step is to collect terms (step 3 from the above procedure).

Example 10.9

Solve the equation,

$2n + 3 = 5n - 1$

Answer

First, collect terms.

$2n - 5n = - 1 - 3$ OK That needs explaining!

What I did here was to shift terms containing n to the left of the equals sign, and shift anything not containing n to the right, **all in one go.** You just consider each term, one at a time and decide which side you want it. Don't worry about 'losing' a term, each term must end up somewhere and you can easily see if one or more is missing. Actually, you can break this down into several smaller steps, but do try it all in one go. OK, it looks difficult but, after a bit of practice, you may surprise yourself.

Let's start again with:

$2n + 3 = 5n - 1$ Shift the 5n to the left and the + 3 to the right.

Firstly just write down what is **already** on the left that you want to **stay** there (in this case, any term containing n) together with the sign (plus or minus) which immediately precedes it. In this case, you'd just then have written down:

$2n$

Now bring over anything that you need to from the right hand side, to the left hand side, reversing the sign as you do so. In this case, we need to bring the (+) 5n over to the left, so it will become - 5n. Now write down the equals sign. This will leave you with:

$2n - 5n =$

Now write down what was **already** on the right hand side, and that you want to **stay** there. **Imagine that this is actually attached and stuck to the equals sign.** This leaves you with:

$2n - 5n = - 1$

Now bring over from the left, what you had ignored earlier, that is the + 3, remembering to change the sign to - 3 as you do so. That's it; done. You now have:

$2n - 5n = - 1 - 3$ Now, returning to step 4 of the procedure, simplify:

$-3n = - 4$ Now, change the signs. What we'll be doing here is effectively to multiply each side by - 1 (you can multiply both sides by negative numbers as well as positives).

$3n = 4$ Now divide by the coefficient of n (i.e. divide by 3)

$n = \frac{4}{3}$ So the solution is $n = \frac{4}{3}$ (i.e. 1 1/3)

Example 10.10

Solve the equation,

$1 - 2n = 4 - 7n$

Answer

First, collect terms.

$-2n + 7n = 4 - 1$ Now simplify.

$5n = 3$ Finally, divide by the 5.

$n = 3/5$

Exercise 10a

Solve the following equations:

1) $4n - 5 = 6$

2) $3n - 4 = n$

3) $10n = 15 - 4n$

4) $3 + 4n = n + 1$

5) $2 - 3n = 8 + 2n$

Now here's an example that starts at step 2 in the procedure (remove brackets).

Example 10.11

Solve the equation,

$$2(n - 3) = 3(2n - 1) + 1$$

Answer

First, expand the brackets.

$2n - 6 = 6n - 3 + 1$	Now, collect terms.
$2n - 6n = - 3 + 1 + 6$	Now simplify.
$-4n = 4$	Now change the signs.
$4n = - 4$	Now divide by the 4.

$$n = -\frac{4}{4} = -1$$

Exercise 10b

Solve the equations:

1) $2(n + 1) = 8$

2) $5(m - 2) = 9$

3) $3(r - 2) = 4(2 - r)$

4) $3p = 5(9 - p)$

* 5) $5(y + 2) - 3(y - 5) = 29$

Just be careful of the minus in front of the bracket and you should get this.

Equations With Denominators

As the original procedure for solving equations states, if an equation contains denominators, that is, if it has fractions in it, the first step should generally be to remove those denominators.

Firstly though, it is important to appreciate, for example, that:

$\frac{1}{2}$ n, or $\frac{1}{2}$ n is the same as $\frac{n}{2}$ — because a half of n is the same as n divided by two.

and that:

$\frac{3}{4}$ n, or $\frac{3}{4}$ n is the same as $\frac{3n}{4}$ — because three quarters of n is the same as n times three, divided by four.

There are advantages and disadvantages of each format, and you should be able to switch back and forth between them, but when solving equations, always convert something of the form ½ n or ¾ n, etc., to the forms $\frac{n}{2}$ and $\frac{3n}{4}$.

Next, I need to remind you a bit about multiplying fractions.

In brief, when multiplying fractions, you 'multiply the tops and multiply the bottoms'. For example:

$$\frac{2}{3} \times \frac{4}{5} = \frac{2 \times 4}{3 \times 5} = \frac{8}{15}$$

or, more generally,

$$\frac{a}{b} \times \frac{c}{d} = \frac{a \times c}{b \times d} = \frac{ac}{bd}$$

It follows that when you multiply a fraction by its denominator (no matter how complex it may be) you are simply left with the numerator (the top part of the fraction). So, for example:

$$\frac{n}{2} \times 2 = \frac{n}{2} \times \frac{2}{1} = \frac{2n}{2} = n$$

Similarly,

$$\frac{2n}{3} \times 3 = 2n$$

and,

$$\frac{ab}{c} \times c = ab$$

and even,

$$\frac{3(a + b)}{(p - q)} \times (p - q) = 3(a + b) \text{ etc.}$$

But, when you multiply a fraction by a number that is not the denominator, effectively you just multiply it into the top, leaving the bottom how it was. So,

$$\frac{n}{2} \times 3 = \frac{n}{2} \times \frac{3}{1} = \frac{3n}{2}$$

Similarly,

$$\frac{2n}{3} \times 4 = \frac{8n}{3}$$

and,

$$\frac{ab}{c} \times d = \frac{abd}{c}$$

and even,

$$\frac{3(a + b)}{(p - q)} \times (m - n) = \frac{3(a + b)(m - n)}{(p - q)}$$

That last one looked a bit tricky, but it isn't really. You just need some practice. So here it is. Persevere!

Exercise 10c

Circle 'True' or 'False' for these. **In the case of 'False', also give the correct answer.**

1) True / False $\quad \frac{n}{5} \times 5 = n$

2) True / False $\quad \frac{n}{6} \times 5 = n$

3) True / False $\quad \frac{2n}{5} \times 5 = 2n$

4) True / False $\quad \frac{2n}{6} \times 5 = 2n$

5) True / False $\quad \frac{bc}{d} \times d = bc$

6) True / False $\quad \frac{bc}{d} \times e = bc$

7) True / False $\quad \frac{b}{3} \times 3 = b$

8) True / False $\quad \frac{5b}{4} \times 4 = 20b$

9) True / False $\frac{2c}{7} \times 7 = 2c$

10) True / False $\frac{2(a + b)}{3} \times 3 = 6(a + b)$

Now let's look at some equations with denominators

Examples

Example 10.12

Solve $\frac{n}{4} = 3$

Answer

First, multiply through by 4.

$n = 3 \times 4$ So,

$n = 12$

Example 10.13

Solve: $\frac{n}{4} = 3 - 3n$

Answer

First, multiply through by 4.

$n = 12 - 12n$ Now collect terms.

$n + 12n = 12$ Now simplify.

$13n = 12$ Now divide by 13.

$n = \frac{12}{13}$

*** Example 10.14**

Solve: $\frac{n}{2} + \frac{n}{3} = 1$

Answer

First, multiply through by 2.

$n + \frac{2n}{3} = 2$ Now, multiply through by 3.

$3n + 2n = 6$ Now simplify.

$5n = 6$ Finally, divide.

$n = \frac{6}{5} = 1\ 1/5$

Exercise 10d

Solve the following equations.

1) $\frac{y}{2} = 3$

2) $\frac{m}{3} = 4$

3) $\frac{c}{3} + c = 5$

4) $\frac{a}{2} = \frac{3}{5}$

Multiply through by 2.

5) $\frac{3}{7} = \frac{4}{n}$

Multiply through by 7, then multiply through by n.

* 6) $\frac{n}{5} + \frac{n}{3} = 2$

Multiply through by 5, then multiply through by 3.

** 7) $\frac{n+1}{5} + \frac{n+2}{3} = 2$

OK, it's got **, but it's not that different from the previous question! But watch out for 'careless mistakes'. I suggest you put the 'n + 1' and 'n + 2' in brackets, and keep the brackets until both denominators have gone.

A Special Case

As with other 'special cases' in this chapter, you don't really need to know the following as you can remove denominators, one by one, but this is a situation that crops up quite often and that can save quite a bit of time, so it's worth a look.

Cross Multiplication

Consider the equation,

$\frac{a}{2} = \frac{b}{3}$ If we multiply though by the 2, we get,

$a = \frac{2b}{3}$ If we now multiply through by the 3, we get,

$3a = 2b$

What this tells us is that, when you have a **single** fraction on one side of the equals sign and another **single** fraction on the other side, you can multiply out the denominators in one go and write:

Left numerator times right denominator	**is equal to**	Right numerator times left denominator

That is to say,

If $\frac{a}{b} = \frac{c}{d}$

then $ad = bc$ (or $bc = ad$)

You may find it helpful to draw a cross centred over the equals sign (like a large letter X) and say to yourself, "That times that equals that times that".

Cross multiplication 'works' even if the fractions are quite complex, as long as there is just one fraction line on the left and one on the right.

Examples

Example 10.15

If $\frac{2}{3} = \frac{5}{a}$ then,

$2a = 15$ and $a = \frac{15}{2} = 7\frac{1}{2}$

Example 10.16

If $\frac{n + 2}{3} = \frac{n - 3}{4}$ then,

$4(n + 2) = 3(n - 3)$ Notice that we needed to bring in brackets here.

$4n + 8 = 3n - 9$

$4n - 3n = - 9 - 8$

$n = - 17$

You can also use this when one side is a fraction and the other side is a single number, for example:

Example 10.17

If $\frac{2a + 3}{4} = 5$ then we can write,

$\frac{2a + 3}{4} = \frac{5}{1}$ and so,

$1(2a + 3) = 5 \times 4$ so,

$2a + 3 = 20$

$2a = 20 - 3$

$2a = 17$

$a = 17/2 = 8\ 1/2$

Exercise 10e

Solve the following equations using cross multiplication.

1) $\frac{5}{3} = \frac{7}{b}$

2) $\frac{a + 3}{4} = \frac{a}{3}$

Remember to introduce brackets.

3) $\frac{3c - 1}{4} = 5$

4) $\frac{2d - 3}{4} = \frac{4 - 5d}{3}$

* 5) $\frac{e + 1}{1 - 3e} = 2$

Section 11: INEQUALITIES

What Are Inequalities?

Equations state that the expressions either side of the ' = ' sign are equal. Inequalities state that they are not equal, but that one side is greater than (or either greater than or equal to) the other.

$>$ means "is greater than"	For example, $5 > 2$
$<$ means "is less than"	For example, $1 < 8$
$\geq$ means "is greater than or equal to"	For example, the days in any month ≥ 28
$\leq$ means "is less than or equal to".	For example, the number of days I work a week ≤ 7

To remember this, you may find it helpful to bear in mind that one side of the $>$ or $<$ sign is wider than the other. The expression that is to the wider side is the one that is the greater. The expression that is to the narrower side is the one that is the lesser.

You need to be able to read inequalities both from left to right and from right to left.

Example 11.1

Re-write the following inequality as it would read from right to left.

$2x - 3 < x + 1$ ($2x - 3$ is less than $x + 1$)

Answer

$x + 1 > 2x - 3$ ($x + 1$ is greater than $2x - 3$)

Exercise 11a

1) Fill in the correct inequality sign ($>$, $<$, $\leq$, or $\geq$) in the space indicated.

(a) 2 10

(b) 10 2

(c) -1 3

(d) -2 . . . . - 6

(e) If I roll a normal dice, the
number showing uppermost will be 6

(f) If I roll two normal die and add
the scores together, the total will be 2

Re-write the following inequalities as they would read from right to left.

2) $x \geq y$

3) $3x > 2y + 1$

4) $3b + 4c < 10$

5) $2(x - 1) \leq 4(x + 2)$

Double Inequalities

Consider the notation,

$2 < x < 8$

This can be called a double inequality because it contains two statements in relation to x. These are:

$x > 2$ (reading the first two parts from right to left) and

$x < 8$ (reading the second two parts from left to right).

So, to 'satisfy' this double inequality x may be anything which is greater than 2 and less than 8. This may not seem as satisfactory as the solution of an equation, that is a **fixed** value, but in fact there are situations in which it is very useful. It might be essential to know that a plant, for example, should not be kept in an environment that is, say, less than 3 degrees centigrade or more than 23 degrees centigrade.

Number Lines

You can represent one or more inequalities on a diagram called a number line. It is generally up to you to decide on how to scale the diagram.

Example 11.2

Represent $x \leq 3$ in a number line diagram.

Answer

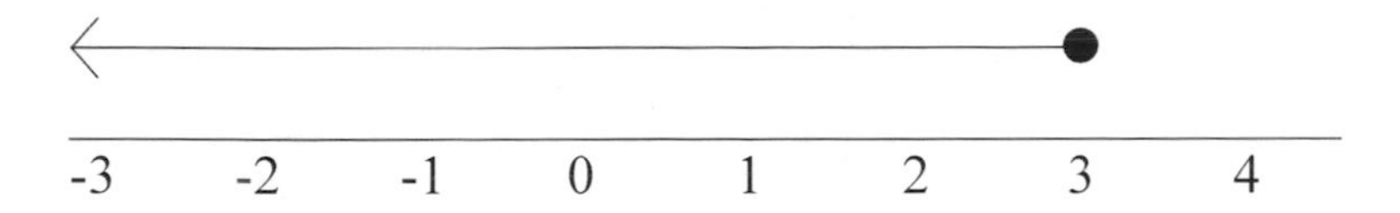

Notice that a solid, filled in circle is used to indicate that the number 3 <u>is</u> included in the range which 'satisfies' the inequality. A hollow circle is used to indicate that the number at that point is <u>not</u> included in the range which satisfies the inequality. The arrow indicates that all other numbers in that direction also satisfy the inequality.

Example 11.3

Represent $x > -1$ in a line diagram.

Answer

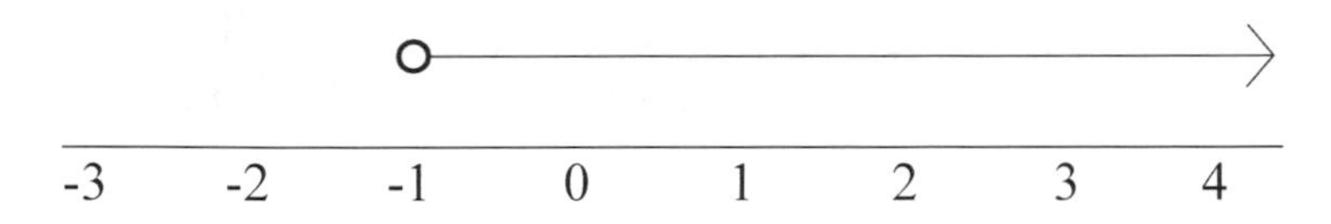

Example 11.4

Draw a number line to represent the values that satisfy the double inequality:

$-1 < x \leq 3$

Answer

This double inequality states that

$x > -1$ (x is greater than - 1) and

$x \leq 3$ (x is less than or equal to 3)

So, to satisfy both, the number line is:

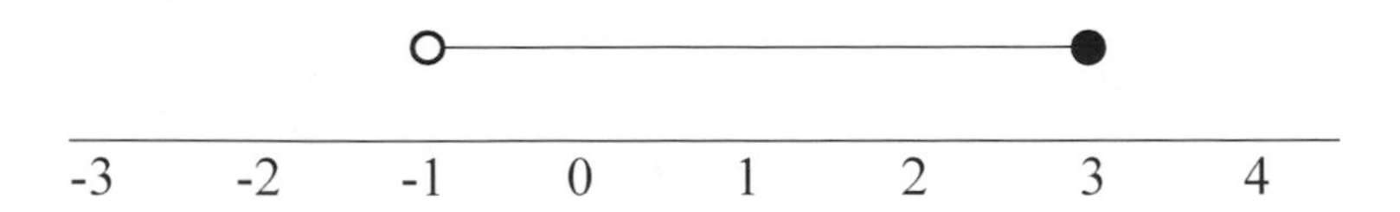

Exercise 11b

Represent the following inequalities using number lines. Pay particular attention to whether to use hollow or filled-in circles at the critical points and/or arrows.

1) $x > 2$

2) $x \leq 3$

3) $x < 2$ and
$x \geq -3$

4) $0 \leq x < 5$

Sometimes you may be asked to list the 'integers' which satisfy one or more inequality. An integer is any whole number (including zero). This can be done using a number line, but it is generally quite straightforward simply to list the whole numbers that fall within the appropriate range.

Example 11.5

State all of the integers which satisfy both inequalities:

$x > -2$ and
$x \leq 3$

Answer

To satisfy both inequalities, x must be greater than - 2 and less than or equal to 3.

The integer solutions are thus - 1, 0, 1, 2 and 3 (remember, 0 is counted as an integer).

Example 11.6

List all integers which satisfy the double inequality:

$-10 \leq x < -3$

Answer

To satisfy the double inequality, x must be greater than or equal to - 10 (reading the left and middle part from right to left) and less than - 3 (reading the middle and right parts).

The integers are thus - 10, - 9, - 8, - 7, - 6, - 5, and - 4.

Exercise 11c

List the integers which satisfy the sets of inequalities:

1) $x < 3$ and
$x \geq -1$

2) $10 \leq x < 15$

3) $x \geq -1$ and
$x < 4$

4) $2 \leq x$ and
$8 > x$

Solving Inequalities

This is much the same as solving ordinary (linear) equations, except for one thing:

When multiplying (or dividing) both sides of an inequality by a negative number, you must change the direction of the inequality.

For example,

$3 < 10$ is a true statement.

If we multiply both sides by, say, 2 we get

$6 < 20$ which is also true.

But if we then multiply each side of the result - 2 it no longer leaves a true statement unless we change the direction of the inequality. That is,

$-12 < -40$ **is not** a true statement, but, switching the inequality:

$-12 > -40$ **is** a true statement.

Thus, if: $-x \leq 3$

Then, multiplying both sides by - 1, we get:

$x \geq -3$.

This can also be verified by algebraic manipulation. If

$$-x \leq 3$$

then moving the x to the right and the number to the left gives,

$$-3 \leq x$$

Then, reading this from right to left gives us:

$$x \geq -3$$

But it is really rather easier just to remember, "If you change the signs you must change the direction" (of the inequality).

Example 11.7

Solve the inequality: $2(x - 3) \geq 5(x + 2)$

Answer

Multiplying out the brackets,

$2x - 6 \geq 5x + 10$, so

$$2x - 5x \geq 10 + 6$$

$$-3x \geq 16$$

$$3x \leq -16$$

Note that we changed the inequality direction because we multiplied both sides by - 1.

$$x \leq -16/3$$

Solution: $x \leq -5\ 1/3$

Example 11.8

Solve the following inequality and represent the solution on a number line:

$$2x + 3 < \frac{3x - 4}{3}$$

Answer

Multiplying both sides by 3 gives,

$6x + 9 < 3x - 4$, so

$6x - 3x < -4 - 9$

$3x < -13$

$x < -13/3$

Solution: $x < -4\ 1/3$

Number line:

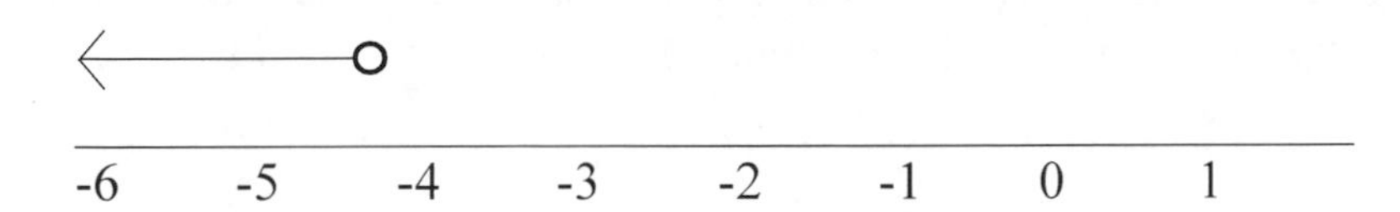

Note especially that the hollow circle lies to the **left** of - 4, not to it's right. This is because minus 4 and 1/3 is more negative than - 4. That is, minus 4 1/3 is less than -4. This is actually quite a subtle but very important thing about numbers. I mention this particularly because, when I first wrote this chapter, I put the hollow circle above the point on the number line above minus 3 and 2/3. No-one's perfect !

Exercise 11d

Solve the following inequalities. Number line solutions are not required unless stated.

1) $2x + 3 \geq 5$

2) $3 - x > 10 + 2x$

3) $3x + 4(5 - x) < 10$

4) Solve and represent in a number line:

$$2 - x < \frac{4 - 5x}{2}$$

Section 12: MAKING EXPRESSIONS & EQUATIONS

Back in section 1, we looked at forming algebraic expressions and equations. Now that you know a bit more, I think it's a good time for some more practice and, where equations are formed, finishing the job off and actually solving them. Don't give up on the questions in this exercise without a fight. As usual, check each answer immediately and do think about the question further if you get an answer wrong.

Note that, providing you use the same units throughout (cm, Kg, minutes, etc., when forming an equation, you do not need to state the units within the equation. For example, you do not need to write,

a centimetres + b centimetres + c centimetres = 10 centimetres

but, providing the units are the same throughout (centimetres here) you can just state:

$a + b + c = 10$

Exercise 12

1) If a boy is n years old now, give an expression, in terms of n, for:

   (a) His age five years ago.

   (b) His age in ten years from now.

2) (a) Find an expression, in pence, for the cost of 3 pencils at p pence each.

   (b) Find an expression, in pence, for 5 ball point pens at b pence each.

   (c) Find an expression, in pence, for the cost of 3 pencils at p pence each and 5 ball point pens at b pence each.

   (* d) Find an expression, in **pounds**, for the cost of 3 pencils at p pence each and 5 ball point pens at b pence each.

3) In a particular week, a man works for n hours on each weekday and for p hours on each of Saturday and Sunday. Form an expression for the total number of hours he works in that week.

4) Find an expression, in centimetres, for the perimeter (distance all the way round) of a rectangle whose length is p cm and whose width is q cm.

5) A man, Alf, has £a. Another man, Bert, has £b. If Alf gives Bert £n, how much will each man then have then?

Alf:

Bert:

6) I think of a number. Let's call it n.

(a) If I subtract 8 from it, what do I get?

(b) If I multiply the result of (a) by 3, what do I then get? (tip: use a bracket)

(c) If, after doing (a) and (b) I get 21, form an equation in n.

(d) Solve this equation to find n, the number I first thought of (and check that this is correct).

7) A room is w metres wide. Its length is 3 metres greater than its width.

(a) In terms of w, what is the length of the room?

(b) If the perimeter (distance all the way round) is 126 metres, form an equation in w.

(c) Solve the equation to find the width of the room.

(d) State the length of the room and confirm that the perimeter is indeed 126 metres.

8) The first of three consecutive numbers is n. In terms of n, what is:

(a) The second number?

(b) The third number?

(c) The sum of the three numbers together?

If the sum of the three numbers is 48:

(d) Form an equation in n.

(e) Solve this equation.

(f) What are the three numbers?

9) The sides of a triangle, in metres, are d, (d - 4) and (d + 2). If the perimeter (you should know what a perimeter is by now!) is 46 cm, form an equation in d, solve it and state the lengths of each side.

10) The angles of a particular triangle are 50°, (n - 20)° and (2n + 30)°. Given that the sum of the angles of any triangle is 180 degrees, form an equation in n, solve it and state the size of each angle.

TEST 4: Time allowed: about 30 minutes.

Solve the following equations:

1) $5n = 20$

2) $\frac{n}{7} = 2$

3) $6n + 5 = 23$

4) $2n - 3 = 8 - 2n$

5) $6(4 - n) = 9$

6) $5(n + 2) + 3(n - 5) = n$

7) $2(3 - n) = 3(n - 2)$

8) $\frac{n}{2} = 3 - n$

9) $\frac{2d - 1}{5} = \frac{1 - 5d}{4}$ **<u>NB 3 marks</u>**

* 10) $\frac{n+1}{2} + \frac{n+2}{3} = 1$

Notice that you cannot cross multiply here, because you have two fractions on the left. You must therefore multiply through by the 2, and then by the 3 in turn. Remember to introduce brackets

NB 4 marks

* 11) My cat is n years old.

(a) If my dog is 2 years older than my cat, give an expression for my dog's age, in terms of n.

(b) If my parrot is twice the age of my dog, give an expression for my parrot's age, in terms of n.

(c) Together, the ages of all three pets come to 18. Form an equation in n and solve it.

(d) State the age of each pet.

12) Solve the inequalities:

(a) $2x + 3 < 11$

(b) $\frac{x+3}{2} \geq 6$

13) Extended question. This may look complicated, but it's not. Just follow the instructions though and don't give up without a fight. If necessary, refer to the answer at the back and then try to continue on your own.

The following is another type of double inequality.

$2x - 20 < 4 - x \leq x - 6$

(a) The first inequality consists of the part on the left, the first inequality sign and the part in the middle. Write this down and solve it.

(b) The second inequality consists of the part in the middle, the second inequality sign and the part on the right. Write this down and solve it.

(c) Draw a number line diagram to represent the values of x which satisfy both inequalities.

(d) List the integers which satisfy both inequalities.

Section 13: REARRANGING FORMULAS

Consider the formula,

$$b = 2a + 3$$

The formula means that to get the value of b, we need to double the value of a and then add on 3. In this case, b is referred to as the **subject** of the formula and we say that the formula **'expresses b in terms of a'**.

Sometimes we may wish to rearrange such a formula, so that 'a' is the subject. In this case, we would get,

$$a = \frac{b - 3}{2}$$ But how?

Rearranging formulas is actually a very similar process to solving equations. Always do it this way, to start with at least. Note that some of the steps listed below will just be omitted, if they aren't relevant. For example, you obviously won't have to remove denominators if there aren't any!

Procedure For Rearranging Formulas

1) Remove denominators (multiply by each in turn or cross multiply).

2) Remove brackets (i.e. expand or multiply them out).

3) If the thing you want to make the subject (let's say it's 'n') occurs on the right hand side of the formula and **not at all** on the left, then just flip the formula around. For example, if you want to make n the subject of the formula $b = 2n - c$, re-write this as $2n - c = b$.

4) Collect terms. If an expression contains the thing we want to make the subject, n, shift it to the left hand side of the equals sign. If an expression doesn't contain n, shift it to the right hand side.

5) Simplify, where possible.

5) Change (all) the signs, if necessary. It is 'necessary' if you have (all) negative ns on the left.

6) This is the bit that everyone forgets (don't think you won't too!). **Take out the common factor if there is one.** Note that, if there is one, it should be the thing that you're making the subject).

7) Divide by the number which precedes n (or by the bracket, if you have taken out a common factor).

8) You may need to do a bit of 'tidying up' at this point but, generally, you should now have n = ... , or something very close to it at this stage.

Just in case you're interested, step 6 is really rather clever. If n appears more than once and can't just be simplified to one term, then what this procedure does is to collect the terms which contain n, so that we **deliberately** created a common factor (of n). We can then divide by the bracket to leave n on its own. If you can appreciate the subtlety of this, you are getting there.

Example 13.1

Make 'a' the subject of the formula:

$b = 2a + 3$

Answer

First, flip it around.

$2a + 3 = b$	Now collect terms.
$2a = b - 3$	No simplifying etc. so just divide.
$a = \frac{b - 3}{2}$	All done!

Example 13.2

Make p the subject of the formula:

$3(1 - p) = 4(p - q)$

Answer

$3(1 - p) = 4(p - q)$	First, remove brackets.
$3 - 3p = 4p - 4q$	Now collect terms.
$-3p - 4p = - 4q - 3$	Now simplify.
$-7p = - 4q - 3$	Now change **all** the signs.
$7p = 4q + 3$	Now divide by 7.
$p = \frac{4q + 3}{7}$	

*** Example 13.3**

Make r the subject of the formula:

$$q = ar + \frac{r}{3}$$

Answer

Firstly, the rs are all on the right, so flip it around.

$ar + \frac{r}{3} = q$ Now, multiply through by 3.

$3ar + r = 3q$ The terms are already collected, so **now, take out the common factor.**

$r(3a + 1) = 3q$ Finally, divide by the bracket.

$$r = \frac{3q}{3a+1}$$

Exercise 13

Rearrange the following to make the symbol given in brackets the subject.

1) $pv = c$ (p)

2) $c = \pi d$ (d) Tip: Flip it round first.

3) $v^2 = 2gh$ (h) And flip it again!

4) $c + pr = n$ (p)

5) $y = mx + c$ (x)

6) $2a - 3b = 4$ (a)

7) $2a - 3b = 4$ (b)

8) $2a + 3b = 5c - 6a$ (a)

9) $s = \pi r(r + h)$ (h)

Flip it around, then multiply out the brackets.

10) $h = \frac{abcd}{500}$ (c)

Remove the denominator first.

11) $r = \frac{d}{3} + 15$ (d)

Flip it around, then multiply through by 3.

12) $w = 2 - \frac{q}{r}$ (q)

Multiply through by r first.

* 13) $w = 2 - \frac{q}{r}$ (r)

A common factor will be needed!

* 14) $V = \frac{2 + n}{(n + r)}$ (r)

Multiply through by (n + r).

** 15) $V = \frac{2 + n}{(n + r)}$ (n)

Don't assume you can't do this!

NB 4 MARKS

Section 14: SIMULTANEOUS EQUATIONS

Simultaneous equations are a pair (or more) of equations, **in two unknowns**, that are both true at the same time. The type that you'll come across at first is where both are linear, for example,

$$2a + 3b = 5$$
$$3a + 4b - 7$$

We cannot solve either of these equations on their own. Thc ncarcst you can get is to make a or b the subject of the equation. The first equation can be rearranged to:

$a = \frac{5 - 3b}{2}$ and the second equation can be rearranged to

$$a = \frac{7 - 4b}{3}$$

But neither of these is a 'solution', as a that would state that 'a' is equal to a specified number. For the equations we have been given, there is only one value of a and one value of b which will make both equations (simultaneously) true. Look again.

$$2a + 3b = 5$$
$$3a + 4b = 7$$

In fact, if you substitute $a = 1$ and $b = 1$, you will find that the first equation does work out as 5 and the second as 7. So the solution is:

$a = 1$ and $b = 1$ Notice that the 'solution' involves a value for a and for b, together.

You might get this by 'inspection' (that is by just looking and guessing) or by trial and error, but you could sit there all day in some cases and not get the solution. We need another system!

There are actually a number of different ways of solving simultaneous equations. I will show you the one that I like best for beginners. Some people would argue that this method is difficult. I would actually disagree but, in any case, this method always works and, if you remember the system, there are fewer opportunities to make mistakes than with other methods. It's also very good practice for rearranging formulas and cross multiplying.

Procedure For Solving Simultaneous Equations

1) Re-write the equations, side by side, with a vertical line between them. For example,

$2a + 3b = 5$	*	$3a + 4b = 7$
	*	
	*	

2) Make ‘a’ the subject of the first equation, directly below the original.

3) Make ‘a’ the subject of the second equation, directly below the original.

4) It follows that the resulting two expressions in ‘b’ must be equal. Write down the equation which states this.

5) Solve the resulting equation in for b.

6) Substitute this value in any one of the formulas in (1), (2), or (3) (the formula in 2 is generally the best option) to give the solution for ‘a’.

7) State the full solution, for example,

   $a = 1, b = 1.$

The only problem is that it uses quite a lot of space so, in an exam, write small and use space elsewhere, if necessary.

Example 14.1

Solve the simultaneous equations:

$2a + 3b = 5$
$3a + 4y = 7$

Answer

Firstly, write them alongside each other. Then, make ‘a’ the subject of each.

$2a + 3b = 5$	*	$3a + 4b = 7$
	*	
$2a = 5 - 3b$	*	$3a = 7 - 4b$
	*	
$\mathbf{a = \frac{5 - 3b}{2}}$	*	$a = \frac{7 - 4b}{3}$
	*	

Therefore $\frac{5 - 3b}{2} = \frac{7 - 4b}{3}$ Now cross multiply.

$3(5 - 3b) = 2(7 - 4b)$ Now expand.

$15 - 9b = 14 - 8b$ Collect terms.

$-9b + 8b = 14 - 15$ Simplify.

$-b = -1$ Change signs.

$\underline{b = 1}$

Substitute in

$$\mathbf{a = \frac{5 - 3b}{2}}$$

$$a = \frac{5 - 3(1)}{2}$$ Note: Always good to use bracket here.

$$a = \frac{5 - 3}{2} = \frac{2}{2} = 1$$

So, $a = 1$, and the solution is:

$\underline{a = 1}$, $\underline{b = 1}$.

Example 14.2

Solve the simultaneous equations:

$4a - 3b = 8$
$6a + b = 1$

Answer

$4a - 3b = 8$	*	$6a + b = 1$
	*	
$4a = 8 + 3b$	*	$6a = 1 - b$
	*	
$\mathbf{a = \frac{8 + 3b}{4}}$	* *	$a = \frac{1 - b}{6}$

Therefore $\frac{8 + 3b}{4} = \frac{1-b}{6}$

$6(8 + 3b) = 4(1 - b)$

$48 + 18b = 4 - 4b$

$18b + 4b = 4 - 48$

$22b = - 44$

$b = - \frac{44}{22}$

$\underline{b = - 2}$ **Substitute in** $\mathbf{a = \frac{8 + 3b}{4}}$

$a = \dfrac{8 + 3(-2)}{4}$ Remember, a bracket helps here.

$a = \dfrac{8 - 6}{4} = \dfrac{2}{4} = \frac{1}{2}$ So, the solution is: $\underline{a = \frac{1}{2}}$ $\underline{b = -2}$

Exercise 14

Solve the simultaneous equations:

1) $3p + 2q = 7$
$p + q = 3$

*
*
*
*
*
*

2) $4a - 3b = 7$
$3a + 2b = 1$

*
*
*
*
*
*
*

3) $3m - 4n - 5$
$2m - 5n = 8$

** 4) $2c + 3 = 5d$
$3d + 1 = 4c$

Tip: These equations are just jumbled up. Make c (or d) the subject of each, Then continue as normal.

* 4) The sum of two numbers, p and q, is 27 and their difference is 3. p is greater than q. Use this information to form two equations in p and q. Solve the equations to find what p and q are (n.b. you must show all working; zero marks for guessing!).

Section 15: QUADRATIC EQUATIONS

Remember that equations with no x^2, x^3, etc were called **linear** equations? Well, equations that contain x^2, but no higher power, are called **quadratic** equations. When these are plotted on a graph, they form a curve, rather than a straight line (which you get from a linear equation).

Quadratic equations can have two solutions, whereas linear equations only have one. For example, the quadratic equation,

$$x^2 - 5x + 4 = 0$$

has two solutions. They are:

$x = 1$, and $x = 4$.

Check that out by substituting them in the equation to see if the result is true:

$(1)^2 - 5(1) + 4 = 0$. So, $x = 1$ is a solution. Similarly:

$(4)^2 - 5(4) + 4 = 0$. So, $x = 4$ is a solution.

If you substitute any other value for x, the resulting statement will not be true. For example,

$(2)^2 - 5(2) + 4 \neq 0$. So, $x = 2$ is **not** a solution.
($\neq$ means 'is not equal to')

So we need another system to solve quadratic equations. In this case, **factorising** will come to our aid. But first, let's consider something about numbers being multiplied together.

Supposing you know that two numbers, p and q, multiply out to another number, say 10. What can we conclude from the fact:

p times q = 10 ?

If we knew that p and q were whole numbers (integers), then we could say that p and q are either 1 and 10, or 2 and 5. But p and q may not be whole numbers, in which case any of the following pairs could work:

0.5 and 20, or 0.25 and 40, or 0.1 and 100, or - 2.5 and - 4, etc.

In fact, the list of possible solutions is endless. So the answer is that we can conclude nothing further at all if we know that p times q is 10.

But, supposing we know that:

p times q = 0. Can you conclude any more?

Think about it yourself a bit, before moving on. Supposing p is 2, what would q have to be, so that they multiply out to 0? Now suppose that p was - 13.76, what would q have to be, so that they multiply out to 0? Cover up the rest of the page and think about it, before I tell you.

The answer in both cases (and every case) is that q would have to be zero. In other words,

If p times q equals 0, then

either p = 0 or q = 0 (or both are 0).

In fact, if a whole string of numbers all multiplied together come to zero, then one or more of the numbers must be 0. For example, if

abcd = 0, then at least one of a, b, c and d must be 0.

This doesn't seem much of a big deal at first, but it is! Supposing:

b(b - 2) = 0 meaning b times (b - 2) = 0,

then either:

b = 0	or
b - 2 = 0	in which case, b = 2

So, the solutions of the equation b(b - 2) = 0 are:

b = 0 and b = 2

Similarly, supposing:

(c + 1)(c - 3) = 0 meaning (c + 1) times (c - 3) = 0

then either:

c + 1 = 0 in which case, c = - 1

or:

c - 3 = 0 in which case, c = 3.

So, the solutions of the equation (c + 1)(c - 3) = 0 are:

c = - 1 and c = 3.

What this means is that, if we have a quadratic equation (or indeed a cubic, or higher powered equation) then if, first of all, we arrange it to the form,

something = 0

then, if we can factorise the 'something', it will follow that one or more of the factors must be equal to zero. So we can solve the quadratic equation. So here's the system.

Procedure For Solving Quadratic Equations

1) Rearrange the equation, if necessary, into the form:
   x-squareds first, xs next, numbers last = 0.

2) Factorise the left hand side (look for a common factor first, otherwise double brackets).

3) Equate the factors in turn to 0. Use this exact format:

   So, either . . . . = 0   or   . . . . = 0
   *
   *
   *

   (that is to say, draw a vertical line below the 'or')

4) Solve the resulting linear equations.

5) State the solutions.

Examples

Solve the following equations:

Example 15.1

$x(x - 2) = 0$

Answer

No need to rearrange or factorise, so straight into,

either $\underline{x = 0}$ or $x - 2 = 0$
* $\underline{x = 2}$
*

The solutions are: $x = 0$ and $x = 2$.

OK, this doesn't say unequivocally that x is 2, for example, but we know that, wherever the equation came from, it's solution is either 2 or zero, or both. In practical situations, we can usually look back to the original problem and conclude which of the alternatives (x is 0 or x is 2) is the more likely. Very often, in fact though, we find that both possibilities are meaningful.

Example 15.2

$(x - 3)(x + 1) = 0$

Answer

No need to rearrange or factorise, so straight into,

So, either $x - 3 = 0$ or $x + 1 = 0$

*

$\underline{x = 3}$ * $\underline{x = -1}$

*

The solutions are: $x = 3$ and $x = -1$.

Example 15.3

$x^2 + 5x + 6 = 0$

Answer

No need to rearrange, so factorise,

$(x + 3)(x + 2) = 0$

So, either $x + 3 = 0$ or $x + 2 = 0$

*

$\underline{x = -3}$ * $\underline{x = -2}$

*

The solutions are: $x = -3$ and $x = -2$.

Example 15.4

$x^2 = 3x$

Answer

First, rearrange,

$x^2 - 3x = 0$ Next, factorise,

$x(x - 3) = 0$

So, either $\underline{x = 0}$ or $x - 3 = 0$

*

* $\underline{x = 3}$

*

The solutions are: $x = 0$ and $x = 3$.

Example 15.5

$x^2 = 3x + 10$

Answer

First, rearrange,

$x^2 - 3x - 10 = 0$ Next, factorise,

$(x - 5)(x + 2) = 0$

So, either $x - 5 = 0$ or $x + 2 = 0$

*

$\underline{x = 5}$ * $\underline{x = -2}$

*

The solutions are: $x = 5$ and $x = -2$.

Exercise 15

Solve the following equations:

1) $n(n - 1) = 0$ No need to expand, just go straight into, either . . . or . . .

2) $(d - 2)(d + 3) = 0$ Same as above.

3) $m^2 + 9m + 20 = 0$ First, factorise

4) $x^2 + 4x = 32$ First rearrange, then factorise.

5) $b^2 = 6b - 9$ First rearrange, then . . . .

* 6) $x = 72 - x^2$

No more clues, you're on your own!

7) $p^2 = 5p$

8) $x^2 - 25 = 0$

* 9) $6m^2 = 15m$

* 10) $(x - 2)(x - 3) = 12$

For the remaining questions, first construct an appropriate equation from the information given. Then solve it. Then look again at the question.

* 11) A number, added to the square of itself, makes 42. If the number is n, how would you write the square of the number? Now write out the equation which says that the number and its square add up to 42. Solve this equation to find out the **two** possible values of n (one is positive and the other is negative). Finally, verify that the solutions work.

* 12) The length of a rectangular room is 2 metres more than its width. If its width is d metres, what is its length, in terms of d? Draw a diagram of a rectangle and label it to represent this. The floor area of the kitchen is 80 square metres. Form an equation expressing these facts. Solve the equation. Then state the width and length of the kitchen. Show all working; no marks for guessing the answer. (The area of a rectangle is calculated by multiplying its length by its width.)

TEST 5 Time allowed: about 30 minutes.

1) Make n the subject of the following formulas:

(a) $a = 5n - 3$

(b) $2n + 3m = 4p$

(c) $2(3n - 4q) = 3(1 - n)$

(* d) $y + \frac{1}{n} = 2$

2) Solve the simultaneous equations:

(a) $5a - 3b = 8$
$6a + 2b = 4$

(* b) $2m + n = 1$
$10m - 3n = 1$

3) Solve, by factors, the equations:

(a) $x^2 = 10x$

(b) $x^2 - 16 = 0$

(c) $x^2 - x = 6$

(d) $x^2 + 52x + 100 = 0$

TEST 6 Revision Time allowed: about 90 minutes.

1) My dog is n years old. My cat is x years younger than my dog. My parrot is x years older than my dog. Write down, in terms of n and / or x,

(a) How old is my cat?

(b) How old is my parrot?

(c) By how much is my parrot older than my cat?

2) If a = - 1, b = - 2 and c = 3, find the values of,

(a) $2a + 4c$

(b) $2ab$

(c) $2ab - 3bc$

(d) b^2c

(e) bc^2

(f) $(bc)^2$

3) Simplify as far as possible,

(a) $p + q + p + p + q$

(b) $3r^3 + 4r^3$

(c) $3mn^2 - 4m^2n + 4mn^2 - m^2n$

4) Simplify,

(a) $p \times q \times p \times p \times q$ (where 'x' means 'times')

(b) $3r^3 \times 4r^3$

(c) $3a \times 4b$

(d) $(-2pq) \times 4p$

(e) $(-2a^2b) \times (-3ab^3)$

5) Express the following in their simplest form,

(a) $3p \div 4q$

(b) $3ab^2 \div ab$

(* c) $\frac{9c^4d^2 \times 4cd^3}{12c^2d}$

6) Expand and simplify, where appropriate,

(a) $3(2a + 4b)$

(b) $-2(3a + 2b - 3c)$

(c) $2ab(3b - 4a^2)$

(d) $3(2a - 3b) + 2(a - 5b)$

(* e) $3(2a - 3b) - 2(a - 5b)$

7) Expand and simplify, where appropriate,

(a) $(a + 3)(a + 2)$

(b) $(a - 3)(a - 2)$

(c) $(2a + 3)(3a - 1)$

(d) $(2a + 3b)(3a - 2b)$

(* e) $(x - 2y)^2$

8) Factorise fully,

(a) $15a + 10b - 5c$

(b) $6b^2 - 2b$

(c) $18a^2b^3 + 12a^3b$

(d) $c^2 + 5c + 6$

(e) $d^2 - 8d + 16$

(f) $x^2 - 1$

(* g) $25x^2 - 16$

9) Solve the equations,

(a) $2p - 6 = 1$

(b) $3(n - 1) = 10$

(c $3x = 4 - 5x$

(d) $2(d - 2) + 3(d + 1) = - 4$

(e) $\frac{n}{3} = 9 - n$ (first multiply through by 3)

(f) $\frac{3b + 2}{2} = \frac{b - 3}{3}$ (cross multiply)

10) The length of the side of a square is x cms. A rectangle is 3 cms longer than the side of the square and its width is 2 cms less than the side of the square. The areas of the square and the rectangle are equal. Draw a square and a rectangle to represent this.

(a) Find an expression, in cms, in terms of x, for the length of the rectangle.

(b) Find an expression, in cms, in terms of x for the width of the rectangle.

(c) Form an equation which states that the areas are equal.

(d) Solve the equation.

(e) State the length of the side of the square, the length of the rectangle and the width of the rectangle.

11) Make x the subject of the following formulas.

(a) $pq = 3xy$

(b) $3y + 4x = 5$

(c) $2(x - a) = 3(b - x)$

(d) $ax + bx = cx + d$

(e) $y = 2 + \frac{3}{x}$

12) Solve the simultaneous equations,

$2a + b = 1$
$4a - 3b = 12$

13) Solve the quadratic equations,

(a) $x^2 - 3x - 4 = 0$

(b) $x^2 + 5x = 0$

(c) $2x^2 - 3x = 0$

(d) $(x - 2)(x + 7) = 0$

(e) $x^2 + 6 = 7x$

(f) $x^2 - 16 = 0$

14) Show that a solution of the equation,

$x^3 - 5x = 50$

lies between $x = 4$ and $x = 5$. Then go on to find this solution, correct to 1 decimal place, showing all your working.

15 (a) Write down <u>two</u> inequalities which together are represented by the number line:

-2 -3 -2 -1 0 1 2 3

(b) List the integers which satisfy,

$-20 \leq x < -15$

(c) List the non-zero even integers which satisfy,

$-1 < x \leq 4$

(d) Solve the inequality,

$2(x - 3) \geq 5(2 + x)$

* 16) (a) Express in algebraic form,

Twice a number, p, plus three times a number, q, is less than three times p plus two times q.

(b) If the above statement is true, which is greater, p or q? You must show your working.

ANSWERS

Test 1 (answers at end)

Exercise 1

1) $6p$
2) $4y + 5$
3) $2n - c$
4) $5abc$
5) $0.5d$, or $\frac{1}{2} d$, or $\frac{d}{2}$, or $d \div 2$.
6) $3p - 4q$
7) $4(a + b)$
8) $\frac{a - b}{3}$
9) $\frac{c + d}{g}$
10) $(c + d)(c + d)$, or $(c + d)^2$
11) $2(a + b) - 3(a - b)$
12) $3a + 4b = 10$
13) $a + b + c = 180$
14) $4(a + b) = 100$
15) $\frac{a + b + c + d}{4} = 8$

Exercise 2

1) 2
2) 11
3) 30
4) 8
5) - 8
6) - 10
7) - ¾
8) - 30
9) 2
10) 1 7/10

Exercise 3

1) 4
2) 81
3) 94
4) 3
5) 16
6) 8

Exercise 4

1) $x = 2$ gives 12. $x = 3$ gives 33. So a solution lies between 2 and 3. $x = 2.7$ (correct to 1d.p.). Note: 2.70 is wrong!
2) $x = 4.4$ (correct to 1d.p.). Note: 4.40 is wrong!

Exercise 5

1) $3a - 2b$
2) $7b$
3) $-9c$
4) d
5) $8b$
6) $10c$
7) $3pq$
8) $11xy$
9) $-2r^2$
10) $3m - n$
11) $x^2 - 2x - 2$
12) $12ab^2 - 3ab + 7$
13) $6b^4 + ab^3 + 5a^2b - 2ab$
14) $4a + 1.5b$
15) $14p^{27}$

Exercise 6

1) $12yz$
2) $6ab$
3) $20m$
4) $8qs$
5) $-yz$
6) $12np$
7) $-18de$
8) $9a^2$
9) $-20et$
10) $-24pqr$
11) $60u^2vw$
12) $-b^2$
13) b^2
14) $6ab^2c$
15) $-16m^4n^5$
16) $-12ab^3$
17) $10q^5r^6$
18) $6a^2q^2$
19) $-pq^4r^3$
20) $-27m^9$

Exercise 7

1) $2p$
2) $-\frac{5a}{6b}$
3) $-\frac{4a}{7b}$
4) $\frac{s}{t}$
5) 2
6) $2b$
7) $4by$
8) $4c$
9) $\frac{4cd}{e}$
10) $-\frac{4cd}{e}$
11) $\frac{4cd}{e}$
12) $\frac{5a^3}{6b^2}$
13) $\frac{a^2p^2}{2b}$
14) $c + 2$
15) $\frac{1}{c+2}$

Exercise 8a

1) $3a + 3$
2) $6e - 8f$
3) $3pq + 4p$
4) $-a - b$
5) $-3a + 2c$
6) $-2p^2 + 2p$
7) $ab + ac - ad$
8) $2a^4 + 6a^3 - 2a^2$
9) $5n + 13$
10) $5a + 1$
11) $6a + 4$
12) $n - 13$
13) $n + 7$
14) $2a - 4b$
15) $3r^2 - 2r + 7$
16) $3p^2 - 2p$
17) $8a$

Exercise 8b

1) $n^2 + 4n + 3$
2) $a^2 + a - 6$
3) $2c^2 + 13c + 15$
4) $5c^2 - 18c - 8$
5) $6a^2 - 13a - 5$
6) $9n^2 + 24n + 16$
7) $a^2 - b^2$
8) $6x^2 + xy - 12y^2$
9) $12x^2 - 11x + 2$
10) $a^2 + b^2 + 2ab + ac + bc$ (in any order)

Test 2 (answers at end)

Exercise 9a

1) $5(a + b)$
2) $3(a - 5)$
3) $2a(2 - b)$
4) $p(d - e)$
5) $3(a + 2b - 3c)$
6) $d(7 - 3d)$
7) $3d(3 - d)$
8) $x(x^2 + x + 1)$
9) $2x(x + 2)$
10) $ab(b - 1)$
11) $ab^2(b - 1)$
12) $mn(m + n)$
13) $3mn(m + 2n)$
14) $3m^4n^2(mn + 2)$
15) $3mn(m + 2n - 3mn)$

Exercise 9b

1) +4 and +5
2) +2 and - 1
3) - 2 and - 3

Exercise 9c

1) $(x + 7)(x + 1)$
2) $(x + 2)(x + 4)$
3) $(x - 2)(x - 5)$
4) $(x - 5)(x - 6)$
5) $(x - 2)(x + 1)$

6) $(x - 5)(x + 3)$
7) $(x + 4)(x - 2)$
8) $(x + 4)(x - 3)$
9) $(x + 4)^2$
10) $(x - 3)^2$
11) $(x + 4)(x - 4)$
12) $(1 + x)(1 - x)$
13) $(2x + 1)(2x - 1)$
14) $(3x + b)(3x - b)$
15) $(x + y)^2$

Exercise 9d

1) $y(y - 4)$
2) $(y + 2)(y - 2)$
3) $(y - 2)^2$
4) $(y - 8)(y - 1)$
5) $2y(y - 2)$
6) $2y^2(1 - 2y)$
7) $(y + z)(y - z)$
8) $y(y^2 + y + 1)$
9) $(5y + 1)(5y - 1)$
10) $(y^2 + 3)^2$

Test 3 (answers at end)

Exercise 10a

1) $2\frac{3}{4}$
2) 2
3) 1 1/14
4) - 2/3
5) - 1 1/5

Exercise 10b

1) 3
2) 3 4/5
3) 2
4) 5 5/8
5) 2

Exercise 10c

1) T
2) F, $\frac{5n}{6}$
3) T
4) F, $\frac{5n}{3}$
5) T
6) F, $\frac{bce}{d}$
7) T
8) F, 5b
9) T
10) F, 2(a + b)

Exercise 10d

1) 6
2) 12
3) 3 3/4
4) 1 1/5
5) 9 1/3
6) 3 3/4
7) 2 1/8

Exercise 10e

1) 4 1/5
2) 9
3) 7
4) 25/26
5) 1/7

Exercise 11a

1) (a) $<$ (d) $>$
(b) $>$ (e) $\leq$
(c) $<$ (f) $\geq$
2) $y \leq x$
3) $2y + 1 < 3x$
4) $10 > 3b + 4c$
5) $4(x + 2) \geq 2(x - 1)$

Exercise 11b

1)

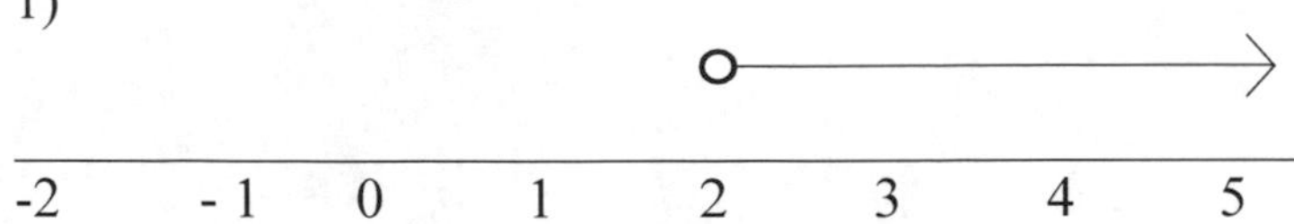

2)

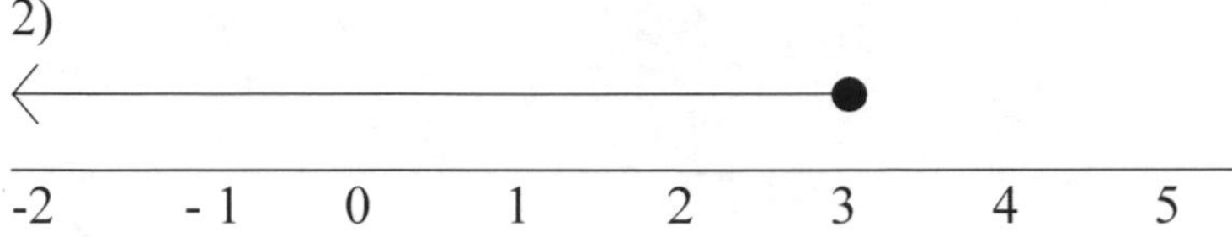

3)

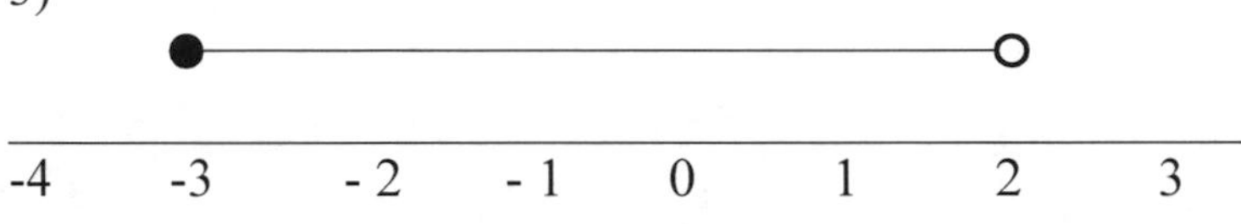

4)

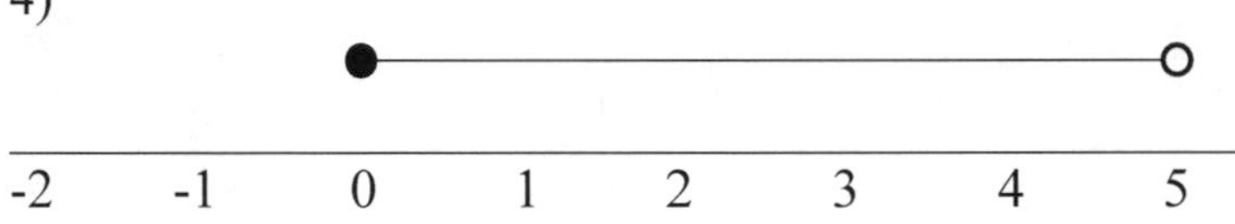

Exercise 11c

1) - 1, 0, 1, 2
2) 10, 11, 12, 13, 14
3) - 1, 0, 1, 2, 3
4) 2, 3, 4, 5, 6, 7

Exercise 11d

1) $x \geq 1$
2) $x < 2\ 1/3$
3) $x > 10$
4) $x < 0$. Note that if $3x < 0$, then $x < \frac{0}{3}$, i.e. $x < 0$.

-3 -2 -1 0 1 2 3 5

Exercise 12

1) (a) n - 5
   (b) n + 10
2) (a) 3p
   (b) 5b

(c) $3p + 5b$

(d) $\frac{3p + 5b}{100}$

3) $5n + 2p$

4) $2p + 2q$, or $2(p + q)$

5) Alf has £(a - n)

Bert has £(b + n)

6) (a) $n - 8$

(b) $3(n - 8)$

(c) $3(n - 8) = 21$

(d) 15

7) (a) $w + 3$

(b) $4w + 6 = 126$

(c) 30 metres

(d) 33 metres, confirmed (33 + 33 + 30 + 30 = 126)

8) (a) $n + 1$

(b) $n + 2$

(c) $3n + 3$

(d) $3n + 3 = 48$

(e) $n = 15$

(f) 15, 16, 17

9) $3d - 2 = 46$. $d = 16$. 16, 12, 18 metres.

10) $3n + 60 = 180$. $n = 40$. 50°, 20°, 110°.

Test 4 (answers at end)

Exercise 13

1) $p = \frac{c}{v}$

2) $d = \frac{c}{\pi}$

3) $h = \frac{v^2}{2g}$

4) $p = \frac{n - c}{r}$

5) $x = \frac{y - c}{m}$

6) $a = \frac{4 + 3b}{2}$

7) $b = \frac{2a - 4}{3}$

8) $a = \frac{5c - 3b}{8}$

9) $h = \frac{s - \pi r^2}{\pi r}$

10) $c = \frac{500h}{abd}$

11) $d = 3r - 45$

12) $q = 2r - wr$

13) $r = \frac{-q}{w-2}$, which is the same as $r = \frac{q}{2-w}$ (** can you explain why?)

14) $r = \frac{2 + n - Vn}{V}$

15) $n = \frac{2 - Vr}{V - 1}$

Exercise 14

1) p = 1, q = 2
2) a = 1, b = - 1
3) m = - 1, n = - 2
4) c = 1, d = 1
5) p + q = 27
p - q = 3
p = 15, q = 12

Exercise 15

1) n = 0 and 1
2) d = 2 and - 3
3) m = - 4 and - 5
4) x = 4 and - 8
5) b = 3 (only)
6) x = - 9 and 8
7) p = 0 and 5
8) x = 5 and - 5
9) m = 0 and 2 ½
10) x = 6 and - 1
11) n^2
$n^2 + n = 42$
n = - 7 and 6
$(-7) + (-7)^2 = 42$
$6 + 6^2 = 42$
12) d + 2
d(d + 2) = 80
d = - 10 or 8. So, d = 8 must be the correct answer, as you cannot have a negative length.
Width = 8m, length = 10 m.

Test 5 (answers at end)

Test 6 Revision Questions (answers at end)

Test 1

1) 8
2) - 4
3) 4
4) - 8
5) 12
6) - 12
7) - 12
8) 12
9) 3
10) - 3
11) - 3
12) 3
13) 14
14) - 14
15) 2
16) 1
17) 9
18) 12
19) 11
20) 4

Test 2

1) T
2) F, - a
3) T
4) T
5) F, - 5a
6) T
7) F, $6a^2$
8) T
9) F, $6a^2$
10) F, 2
11) T
12) T
13) F, 6p
14) T
15) F, $\frac{a + b + c}{d}$
16) T

17) T
18) F, 15
19) F, - 4
20) F, a^5
21) T
22) F, 5a
23) T
24) F, 16
25) F, $13x$
26) F, $3x$
27) T
28) T
29) F, $40a^2$
30) T
31) T
32) F, (not the same)
33) T
34) F, $-x^2 + 2x - 3$
35) F, - 2n
36) T
37) $x = 2$ gives 10 (too small).
$x = 3$ gives 45 (too big). Therefore a solution lies between 2 and 3.
38) 6.6

Test 3

1) $12a^2 - 4a^3$
2) $5a - b$
3) $6a^2 + 7a + 2$
4) $a^2 - 9$
5) $3n^2 - 5n + 2$
6) $4a^2 + 12ab + 9b^2$
7) $2a^2 + 4a$
8) $2a^2b^3(3 - 7ab^2)$
9) $(a - 1)(a - 4)$
10) $a(a - 25)$
11) $(a + 5)(a - 5)$
12) $(a - 10)(a + 9)$
13) $a - 5$

Test 4

1) 4
2) 14
3) 3
4) 2 3/4

5) 2 1/2
6) 5/7
7) 2 2/5
8) 2
9) 3/11
10) - 1/5
11) (a) n + 2
(b) 2(n + 2)
(c) 4n + 6 = 18, n = 3
(d) cat = 3, dog = 5, parrot = 10
12) (a) $x < 4$
(b) $x \geq 9$
13) (a) $2x - 20 < 4 - x$
$x < 8$
(b) $4 - x \leq x - 6$
$x \geq 5$
(c)

4 5 6 7 8 9

(d) 5, 6, 7

Test 5

1) (a) $n = \frac{a+3}{5}$
(b) $n = \frac{4p - 3m}{2}$
(c) $n = \frac{3+8q}{9}$
(d) $n = -\frac{1}{y-2}$, which is the same as $n = \frac{1}{2-y}$
2) (a) a = 1, b = - 1
(b) m = ¼, n = ½
3) (a) $x = 0$ and 10
(b) $x = 4$ and - 4
(c) $x = 3$ and - 2
(d) $x = -2$ and - 50

Test 6 Revision

1) (a) (n - x) years
(b) (n + x) years
(c) $2x$ years

2) (a) 10
(b) 4
(c) 22
(d) 12
(e) - 18
(f) 36

3) (a) $3p + 2q$
(b) $7r^3$
(c) $7mn^2 - 5m^2n$

4) (a) p^3q^2
(b) $12r^6$
(c) $12ab$
(d) $-8p^2q$
(e) $6a^3b^4$

5) (a) $\frac{3p}{4q}$
(b) $3b$
(c) $3c^3d^4$

6) (a) $6a + 12b$
(b) $-6a - 4b + 6c$
(c) $6ab^2 - 8a^3b$
(d) $8a - 19b$
(e) $4a + b$

7) (a) $a^2 + 5a + 6$
(b) $a^2 - 5a + 6$
(c) $6a^2 + 7a - 3$
(d) $6a^2 + 5ab - 6b^2$
(e) $x^2 - 4xy + 4y^2$

8) (a) $5(3a + 2b - c)$
(b) $2b(3b - 1)$
(c) $6a^2b(3b^2 + 2a)$
(d) $(c + 3)(c + 2)$
(e) $(d - 4)^2$
(f) $(x + 1)(x - 1)$
(g) $(5x + 4)(5x - 4)$

9) (a) 3 1/2
(b) 4 1/3
(c) ½
(d) - 3/5
(e) 6 3/4
(f) - 1 5/7

10) (a) $x + 3$
(b) $x - 2$
(c) $x^2 = (x + 3)(x - 2)$
(d) $x = 6$
(e) 6, 9, 4

11) (a) $x = \frac{pq}{3y}$

(b) $x = \frac{5 - 3y}{4}$

(c) $x = \frac{3b + 2a}{5}$

(d) $x = \frac{d}{a+b-c}$

(e) $x = \frac{3}{y-2}$

12) a = 1½, b = - 2

13) (a) x = 4 and - 1

(b) x = 0 and - 5

(c) x = 0 and 1 ½

(d) x = 2 and - 7

(e) x = 6 and 1

(f) x = 4 and - 4

14) x = 4 gives 44 (too small). x = 5 gives 100 (too big).
Therefore a solution lies between x = 4 and x = 5.
x = 4.1 (1dp) (NB The answer, 4.1**0** is wrong.)

15) (a) $x \geq -2$ and $x < 1½$

(b) - 20, - 19, - 18, - 17, - 16

(c) 2, 4

(d) $x \leq 5\ 1/3$

16) (a) 2p + 3q < 3p + 2q

(b) p is greater than q.

About the Author

Will Duncombe grew up in Oxford, England, in the 1960s. After obtaining his degree in psychology at Exeter University, he returned to Oxford, where he was employed at Edward Greene's Tutorial Establishment as an odd job boy. Well, what else can you do with a psychology degree? Later he worked there as an O level maths tutor and then as the Bursar. In 1976, he cofounded St. Andrew's Private Tutorial Centre in Cambridge and, as such, was arguably the youngest ever English headmaster. In 1985, leaving St. Andrew's as a going concern, which still exists today, he and his wife, Fiona, moved to Brighton, on the south coast, and set up a new school, Bartholomews Tutorial College, where he still works. Barts seems to remain a well-kept secret, but it is a place that Will is very proud of and whose purpose is simply to offer complete academic flexibility. Will reckons that everyone should try to write a book in their lifetime. This is his.